Albert William Smith

Elementary machine Design

Albert William Smith

Elementary machine Design

ISBN/EAN: 9783743344303

Manufactured in Europe, USA, Canada, Australia, Japa

Cover: Foto ©ninafisch / pixelio.de

Manufactured and distributed by brebook publishing software (www.brebook.com)

Albert William Smith

Elementary machine Design

ELEMENTARY

MACHINE DESIGN

— BY —

ALBERT W. SMITH

Professor of Mechanical Engineering
Leland Stanford Jr. University

CALIFORNIA
STANFORD UNIVERSITY PRESS
1895

Copyrighted, 1894, by A. W. Smith.

ERRATA FOR ELEMENTARY MACHINE DESIGN.

Page 13, fifth line, for "c" read "d."
" 19, first line, before "The centro ——" insert "Fig. 19."
" 22, twelfth line from bottom, after "diagram," insert "transfer to a 'time base' as on page 33."
" 24, twenty-sixth line, invert the expression "$\dfrac{\text{max. } Vl \text{ of slider}}{Vl \text{ of } ab}$."

Fig. 23, reverse arrow that indicates the direction of rotation.
" 24, indicate the intersection of a and e by "C."

Page 27, first line, for "cd" read "af."
" 30, fifteenth line, for "x" read "z."
" 32, eighteenth line, for "revolutions" read "strokes."
" 40, sixth line from bottom, for "38" read "37."

Fig. 40, indicate arc of small circle between bc and M by "D."

Page 44, twelfth line, for "CE" read "CD."
" 47, first line, for "45" read "44."
" 68, first part of seventh line from bottom, for "1·5" read "3."
" 73, fifteenth line, for "C" read "O."
" 84, third line, for "Sin" read "Tan."
" 86, fifth line, for "sin" read "tan."
" 90, ninth line, for "$T_1 + T_2$" read "$T_1 - T_2$."
" 91, eighth line, after a insert "Fig. 96."
" 98, twentieth line, for "$\dfrac{W'}{g}(\quad)$" read "$\dfrac{W'}{2g}(\quad)$."
" 98, in equations "$W =$ ——" insert "2" as a factor in the numerator of the right-hand member, and make corresponding change in result.
" 158, eighth line, for "D" read "O."

TO

PROFESSOR JOHN E. SWEET

THIS BOOK IS DEDICATED

AS AN EXPRESSION OF AFFECTION, AND IN

ACKNOWLEDGMENT OF YEARS OF HELPFULNESS.

PREFACE.

ONE can never become a machine designer by studying a book. The true designer is one whose judgment is ripened by experience in constructing and operating machines. Mr. William B. Bement, a designer in the true sense, once said that he thought it useful to figure the strength of machine parts, because the results were *suggestive* to the designer. One may know thoroughly the laws which govern the transmission of energy; may understand much concerning the nature of constructive materials ; may know how to obtain results by mathematical processes ; and yet be unable to design a good machine. One needs also to know the thousand and one things connected with practice, which constantly modify design, so that one can take the results of computation and accept, reject, and modify, until the machine will, when constructed, do its required work satisfactorily and enduringly. The writer once heard Professor John E. Sweet say : " It is comparatively easy to design a good *new* machine, but it is very hard to design a machine which will be good when it is old." This quality of foresight only comes with long experience.

There is, however, a certain part of the designer's mental equipment which may be furnished in the class-room, or by books. This is the writer's excuse for the following pages.

Even Elementary Machine Design cannot be treated exhaustively. The kinds of machines are too numerous, and their differences are too great. An effort is made here to *suggest* methods of reasoning, rather than to give rules. A knowledge of the usual university course in pure and applied mathematics is presupposed.

The part upon "Motion in Machines" could not have been written without the use of the excellent book "The Mechanics of Machinery," by Prof. A. B. W. Kennedy. To him and to Prof. L. M. Hoskins, to whom the writer has so often gone for the help which never failed, grateful acknowledgment is here made. In several places acknowledgment is made to others; yet the writer feels that he has failed, though unintentionally, to give credit for much of the best he has received.

A. W. S.

STANFORD UNIVERSITY, CALIFORNIA, January, 1895.

CONTENTS.

	PAGE
PREFACE	v
INTRODUCTION	ix

CHAPTER I.
PRELIMINARY 1

CHAPTER II.
MOTION IN MECHANISMS 11

CHAPTER III.
ENERGY IN MACHINES 28

CHAPTER IV.
PARALLEL OR STRAIGHT LINE MOTIONS . . 37

CHAPTER V.
TOOTHED WHEELS, OR GEARS 39

CHAPTER VI.
CAMS 72

CHAPTER VII.
BELTS 75

CHAPTER VIII.
DESIGN OF FLY-WHEELS 93

CHAPTER IX.
RIVETED JOINTS 100

CONTENTS.

CHAPTER X.
DESIGN OF JOURNALS . . 110

CHAPTER XI.
SLIDING SURFACES 126

CHAPTER XII.
BOLTS AND SCREWS AS MACHINE FASTENINGS . . 130

CHAPTER XIII.
MEANS FOR PREVENTING RELATIVE ROTATION . 137

CHAPTER XIV.
FORM OF PARTS AS DICTATED BY STRESS . 140

CHAPTER XV.
MACHINE SUPPORTS . . . 146

CHAPTER XVI.
MACHINE FRAMES 150

INDEX . . . 161

INTRODUCTION.

In general there are four considerations of prime importance in designing machines: I. Adaptation, II. Strength and Stiffness, III. Economy, IV. Appearance.

I. This requires all complexity to be reduced to its lowest terms in order that the machine shall accomplish the desired result in the most direct way possible, and with greatest convenience to the operator.

II. This requires the machine parts subjected to the action of forces to sustain these forces, not only without rupture, but also without such yielding as would interfere with the accurate action of the machine. In many cases the forces to be resisted may be calculated, and the laws of Mechanics, and the known qualities of constructive materials become factors in determining proportions. In other cases the force, by the use of a "breaking piece," may be limited to a maximum value, which therefore dictates the design. But in many other cases the forces acting are necessarily unknown; and appeal must be made to the precedent of successful practice, or to the judgment of some experienced man, until one's own judgment becomes trustworthy by experience.

In proportioning machine parts, the designer must always be sure that the stress which is the basis of the calculation or the estimate, is the maximum possible stress. Otherwise the part will be incorrectly proportioned. For instance, if the arms of a pulley were to be designed solely on the assumption that they endure only the transverse stress due to the belt tension, they would be found to be absurdly small, because the stresses resulting from the shrinkage of the casting in cooling, are often far greater than those due to the belt pull.

The design of many machines is a result of what may be called "machine evolution." The first machine was built according to the best judgment of its designer; but that judgment was fallible, and some part yielded under the stresses sustained; it was replaced by a new part made stronger; it yielded again, and again was enlarged, or perhaps made of some more suitable material; it then sustained the applied stresses satisfactorily. Some other part yielded too much under stress, although it was entirely safe from actual rupture; this part was then stiffened, and the process continued, till the whole machine became properly proportioned for the resisting of stress. Many valuable lessons have been learned from this process; many excellent machines have resulted from it. There are, however, two objections to it: it is slow and very expensive, and if any part had originally an excess of material, it is not changed; only the parts that yield are perfected.

III. The attainment of economy does not necessarily mean the saving of metal or labor, although it may mean that. To illustrate: Suppose that it is required to design an engine lathe for the market. The competition is sharp; the profits are small. How shall the designer change the design of the lathes on the market to increase profits ? (a) He may, if possible, reduce the weight of metal used, maintaining strength and stiffness by better distribution. But this must not increase labor in the foundry or machine shop, nor reduce weight which prevents undue vibrations. (b) He may design special tools to reduce labor without reduction of the standard of workmanship. The interest on the first cost of these special tools, however, must not exceed the possible gain from increased profits. (c) He may make the lathe more convenient for the workmen. True economy permits some increase in cost to gain this end. It is not meant that elaborate and expensive devices are to be used, such as often come from men of more inventiveness than judgment, and which usually find their level in the scrap heap; but that if the parts can be rearranged, or in any way changed so that the lathes-man shall select this lathe to use because it is handier, when other lathes are available, then economy has been served, even though the cost has been somewhat increased ; because the favorable opinion of intelligent workmen means increased sales.

In (a) economy is served by a reduction of metal ; in (b) by a reduction of labor ; in (c) it may be served by an increase of both labor and material.

The addition of material largely in excess of that necessary for strength and rigidity, to reduce vibrations, may also be in the interest of economy, because it may increase the durability of the machine and its foundation ; may reduce the expense incident upon repairs and delays, thereby bettering the reputation of the machine, and increasing sales.

Suppose, to illustrate further, that a machine part is to be designed, and either of two forms, A or B, will serve equally well. The part is to be of cast iron. The pattern for A will cost twice as much as for B. In the foundry and machine shop, however, A can be produced a very little cheaper than B. Clearly then, if but one machine is to be built, B should be decided on ; whereas, if the machine is to be manufactured in large numbers, A is preferable. Expense for patterns is a first cost. Expense for work in the foundry and machine shop is repeated with each machine.

Economy of operation also needs attention. This depends upon the efficiency of the machine ; *i. e.*, upon the proportion of the energy supplied to the machine which really does useful work. This efficiency is increased by the reduction of useless frictional resistances, by careful attention to the design and means of lubrication, of rubbing surfaces.

In order that economy may be best attained, the machine designer needs to be familiar with all the processes used in the construction of machines — pattern making, foundry work, forging, and the processes of the machine shop — and must have them constantly in mind, so that while each part designed is made strong enough and stiff enough, and properly and conveniently arranged, and of such form as to be satisfactory in appearance, it also is so designed that the cost of construction is a minimum.

IV. The fourth important consideration is Appearance. There is a *beauty* possible of attainment in the design of machines which is always the outgrowth of a *purpose*. Otherwise expressed: A machine to be *beautiful* must be *purposeful*. Ornament for ornament's sake is seldom admissible in machine design. And yet the striving for a pleasing effect is as much a part of the duty of the machine designer as it is a part of the duty of an architect.

ELEMENTARY MACHINE DESIGN.

CHAPTER I.

PRELIMINARY.

1. The solution of problems in machine design involves the consideration of **force, motion, work,** and **energy.** It is assumed that the student understands clearly what is meant by these terms.

A complete cycle of action of a machine is such an interval that all conditions in the machine are the same at its beginning and end.

The law of **Conservation of Energy** underlies every machine problem. This law may be expressed as follows: The sum of energy in the universe is constant. Energy may be transferred in space; it may be changed from one of its several forms to another; but it cannot be created or destroyed.

The application of this law to machines is as follows: A machine receives energy from a source, and uses it to do useful and useless work. During a complete cycle of action of the machine, the energy received equals the total work done. In other words, a machine gives out, in some way, during each cycle, all the energy it receives; but it cannot give out more than it receives; or, considering a cycle of action,

energy received = useful work + useless work.

When any two of these quantities are given, or can be estimated, the third quantity becomes known.

2. **Function of Machines.**— Nature furnishes sources of *energy*, and the supplying of human needs requires *work* to be done. **The function of machines is to cause matter possessing energy to do useful work.**

Illustration.— The water in a mill pond possesses energy (potential) by virtue of its position. The earth exerts an attractive force upon it. If there is no outlet, the earth's attractive force cannot cause motion; and hence, since motion is a necessary factor of work, no work is done.

If the water overflows the dam, the earth's attraction causes that part of it which overflows to *move* to a lower level, and before it can be brought to rest again, it does work against the force which brings it to rest. If this water simply falls upon rocks, its energy is transformed into heat, with no useful result.

But if the water be led from the pond to a lower level, in a closed pipe which connects with a water-wheel, it will exert pressure upon the vanes of the wheel (because of the earth's attraction), and will cause the wheel and its shaft to rotate against resistance, whereby it may do useful work. The water-wheel is a **machine** and is called a **Prime Mover**, because it is the first link in the machine chain between natural energy and useful work.

Since it is usually necessary to do the required work at some distance from the necessary location of the water-wheel, **Machinery of Transmission** is used (shafts, pulleys, belts, cables, etc.), and the rotative energy is rendered available at the required place.

But this rotative energy may not be adapted to do the required work; the rotation may be too slow or too fast; a resistance may need to be overcome in straight, parallel lines, or at periodical intervals. Hence **Machinery of Application** is introduced to transform the energy to meet the requirements of the work to be done. Thus the chain is complete, and the potential energy of the water does the required useful work.

The chain of machines which has the steam boiler and engine for its prime mover, transforms the potential heat energy of fuel into useful work. This might be analyzed in a similar way.

3. Force Opposed by Passive Resistance. — A force may act without being able to produce motion (and hence without being able to do work), as in the case of the water in a mill pond without overflow or outlet. This may be further illustrated: Suppose a force, say hand pressure, to be applied vertically to the top of a table. The material of the table offers a *passive resistance*, and the force is unable to produce motion, or to do work. But if the table top were supported upon springs, the applied force would overcome the elastic resistance of the springs, through a certain space, and would do work. It is therefore possible to offer passive resistance to such forces as may be required not to produce motion; thereby rendering them incapable of doing work.

4. Constrained Motion. — By watching the action of a machine, it is seen that certain definite motions occur, and that any departure from these motions, or the production of any other motions, would result in derangement of the action of the machine. Thus, the spindle of an engine lathe turns accurately about its axis; the cutting tool moves parallel to the spindle's axis; and an accurate cylindrical surface is thereby produced. If there were any departure from these motions, the lathe would fail to do its required work. In all machines certain definite motions must be produced, and all other motions must be prevented; or, in other words, motion in machines must be *constrained*.

In the case of a " free body," acted on by a system of forces, not in equilibrium, motion results in the direction of the resultant of the system. If another force be introduced whose line of action does not coincide with that of the resultant, the direction of the line of action of the resultant is changed, and the body moves in this new direction. The character of the motion, therefore, is dependent upon the forces which produce the motion. This is called **free motion**.

Example. — In Fig. 1, suppose the free body M to be acted on by the concurrent forces $1, 2$, and 3, whose lines of action pass through the center of gravity of M. The line of action of the resultant of these forces is $A B$, and the body's centre of gravity would move along this line.

If another force, 4, be introduced, CD becomes the line of action of the resultant, and the motion of the body is along the line CD.

Constrained motion differs from free motion in being independent of the forces which produce it. If any force, not sufficiently great to produce deformation, be applied to a body whose motion is constrained, the result is either a certain predetermined motion, or no motion at all.

In a machine there must be provision for resisting every possible force which tends to produce any but the required motion. This provision is usually made by means of the *passive resistance* of properly formed and sufficiently resistant metallic surfaces.

Illustration I. — Fig. 2 represents a section of a wood lathe headstock. It is required that the spindle, S, and the attached cone pulley, C, shall have no other motion than rotation about the axis of the spindle. If any other motion is possible, this machine part cannot be used for the required purpose. At A and B the cylindrical surfaces of the spindle are enclosed by accurately fitted bearings, or internal cylindrical surfaces. Suppose any force, P, whose line of action lies in the plane of the paper, to be applied to the cone pulley. It may be resolved into a radial component, R, and a tangential component, T. The passive resistance of the cylindrical surfaces of the journal and its bearing, prevents R from producing motion; while it offers no resistance, friction being disregarded, to the action of T, which is allowed to produce the required motion, *i.e.*, rotation about the spindle's axis. If the line of action of P pass through the axis, its tangential component becomes zero, and no motion results. If the line of action of P become tangential, its radial component becomes zero, and P is wholly applied to produce rotation. If a force Q, whose line of action lies in the plane of the paper, be applied to the cone, it may be resolved into a radial component, N, and a component, M, parallel to the spindle's axis. N is resisted as before by the journal and bearing surfaces, and M is resisted by the shoulder surfaces of the bearings, which fit against the shoulder surfaces of the cone pulley. The force Q can therefore produce no motion at all.

In general, any force applied to the cone pulley may be resolved into a radial, a tangential, and an axial component. Of these only the tangential component is able to produce motion; and that motion is the motion required. The constrainment is therefore complete; *i. e.*, there can be no motion except rotation about the spindle's axis. This result is due to the passive resistance of metallic surfaces.

Illustration II.— R, Fig. 3, represents, with all details omitted, the "ram," or portion of a shaping machine which carries the cutting tool. It is required to produce plane surfaces, and hence the "ram" must have accurate rectilinear motion parallel to HK. Any deviation from such motion renders the machine useless.

Consider **A**. Any force which can be applied to the ram, may be resolved into three components: one vertical, one horizontal and parallel to the paper, and one perpendicular to the paper. The vertical component, if acting upward, is resisted by the plane surfaces in contact at C and D; if acting downward, it is resisted by the plane surfaces in contact at E. Therefore no vertical component can produce motion. The horizontal component parallel to the paper is resisted by the plane surfaces in contact at F or G, according as it acts toward the right or left. The component perpendicular to the paper is free to produce motion parallel to its line of action; but this is the motion required.

Any force, therefore, which has a component perpendicular to the paper, can produce the required motion; but no other motion. The constrainment is therefore complete, and the result is due to the passive resistance offered by metallic surfaces.

Complete Constrainment is not always required in machines. It is only necessary to prevent such motions as interfere with the accomplishment of the desired result.

The **weight** of a moving part is sometimes utilized to produce constrainment in one direction. Thus in a planer table, and in some lathe carriages, downward motion, and unallowable side motion, are resisted by metallic surfaces; while upward motion is resisted by the weight of the moving part.

Since the motions of machine parts are independent of the forces producing them, it follows that the relation of such motions may be determined without bringing force into the consideration.

5. Kinds of Motion in Machines.—Motion in machines may be very complex, but it is chiefly **plane motion**.

When a body moves in such a way that any section of it remains in the same plane, its motion is called plane motion. All sections parallel to the above section must also remain, each in its own plane. If the plane motion is such that all points of the moving body remain at a constant distance from some line, AB, the motion is called **rotation** about the axis AB. *Example.*—A line shaft with attached parts.

If all points of a body move in straight parallel paths, the motion of the body is called **rectilinear translation**. *Examples.*— Engine cross-head, lathe carriage, planer table, shaper ram. Rectilinear translation may be conveniently considered as a special case of rotation, in which the axis of rotation is at an infinite distance, at right angles to the motion.

If a body moves parallel to an axis about which it rotates, the body is said to have **helical** or **screw motion**. *Example.*—A nut turning upon a stationary screw.

If all points of a body, whose motion is not plane motion, move so that their distances from a certain point, O, remain constant, the motion is called **spheric motion**. This is because the points move in the surface of a sphere whose centre is O. *Example.*—The balls of a fly-ball steam-engine governor, when the position of the valve is changing.

6. Relative Motion.—The motion of any machine part, like all known motion, is relative motion. It is studied by reference to some other part of the same machine. Some one part of a machine is usually (though not necessarily) fixed; *i. e.*, it has no motion relatively to the earth. This fixed part is called the **frame** of the machine. The motion of a machine part may be referred to the frame, or, as often necessary, to some other part which also has motion relatively to the frame.

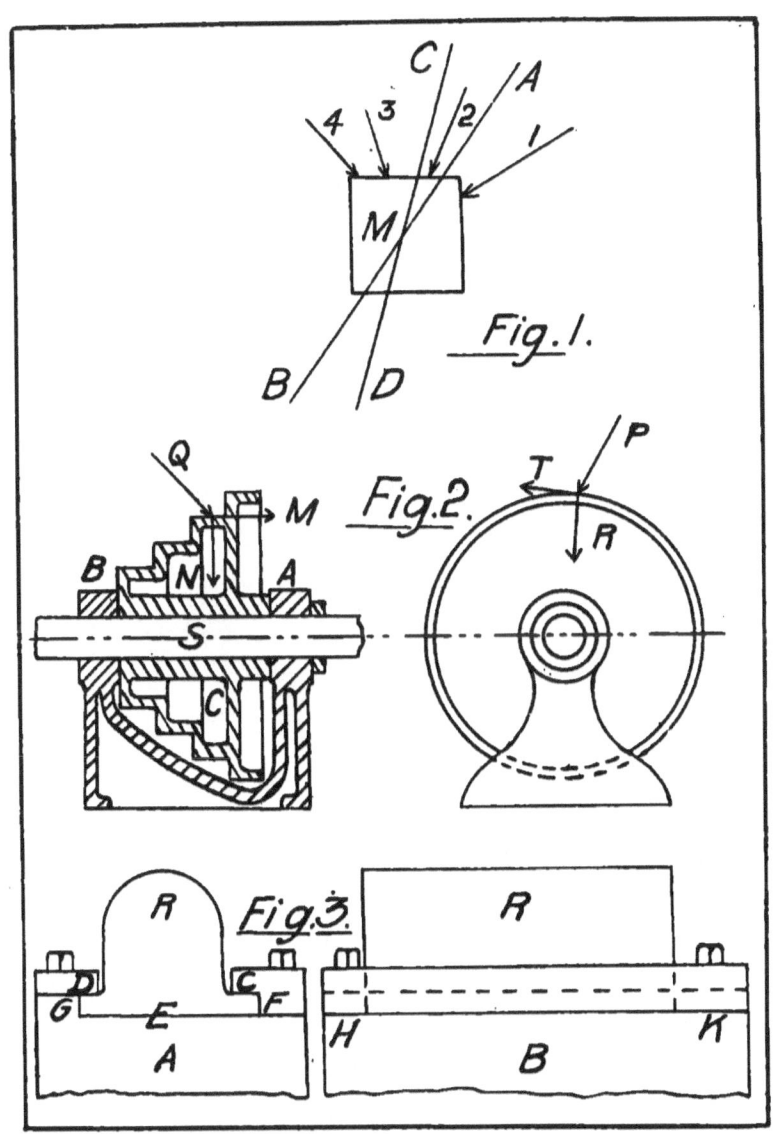

PRELIMINARY. 7

The kind and amount of relative motion of a machine part, depends upon the motions of the part to which its motion is referred.

Illustration. — Fig. 4 shows a press. A is the frame; C is a plate which is so constrained that it may move vertically, but cannot rotate relatively to A. Motion of rotation is communicated to the screw B. The motion of B referred to A is helical motion, *i. e.*, combined rotation and translation. C, however, shares the translation of B, and hence there is left only rotation as the relative motion of B and C. The motion of B referred to C is rotation. The motion of C referred to B is rotation. The motion of C referred to A is translation.

In general, if two machine members, M and N, move relatively to the frame, the relative motion of M referred to N depends on how much of the motion of N is shared by M. If M and N have the same motions relatively to the frame, they have no motion relatively to each other.

Conversely, if two bodies have no relative motion, they have the same motion relatively to a third body. Thus in Fig. 4, if the constrainment of C were such that it could share B's rotation, as well as its translation, then C would have helical motion relatively to the frame, and no motion at all relatively to B. This is assumed to be self-evident.

A **rigid body** is one in which the distance between elementary portions is constant. No body is absolutely rigid, but usually in machine members the departure from rigidity is so slight that it may be neglected.

Many machine members, as springs, etc., are useful because of their lack of rigidity.

Points in a rigid body can have no relative motion, and hence *must all have the same motion*.

7. Instantaneous Motion, and Instantaneous Centres or Centros. — Points of a moving body trace more or less complex paths. If a point be considered as moving from one position in its path to another indefinitely near, its motion is called **instantaneous motion**.

8 MACHINE DESIGN.

The point is moving, for the instant, along a straight line joining the two indefinitely near together positions, and such a line is a tangent to the path. In problems which are solved by the aid of the conception of instantaneous motion, it is only necessary to know the *direction* of motion; hence, for such purposes, *the instantaneous motion of a point is fully defined by a tangent to its path through the point*.

Thus in Fig. 5, if a point is moving in the path APB, when it occupies the position P the tangent TT represents its instantaneous motion. Any number of curves could be drawn tangent to TT at P, and any one of them would be a possible path of the point; but whatever path it is following, its *instantaneous motion* is represented by TT. The instantaneous motion of a point, is therefore independent of the *form* of its path. Any one of the possible paths may be considered as equivalent, for the instant, to a circle whose centre is anywhere in the normal NN.

In general, the instantaneous motion of a point, A, is equivalent to rotation about some point, B, in a line through the point, A, perpendicular to the direction of its instantaneous motion.

Let the instantaneous motion of a point, A, Fig. 6, in a section of a moving body be given by the line TT. Then the motion is equivalent to rotation about some point of the line AB as a centre; but it may be any point, and hence the instantaneous motion of the *body* is not determined. But if the instantaneous motion of another point, C, be given by the line T_1T_1, this motion is equivalent to rotation about some point of CD. But the points A and C are points in a rigid body, and can have no relative motion, and must have the same motion, *i. e.*, rotation about the same centre. A rotates about some point of AB, and C rotates about some point of CD; but they must rotate about the same point, and the only point which is at the same time in both lines, is their intersection, O. Hence A and C, and all other points of the body, rotate for the instant about an axis of which O is the projection; or, in other words, the instantaneous motion of the body is rotation about an axis of which O is the projection. This axis is the *instantaneous*

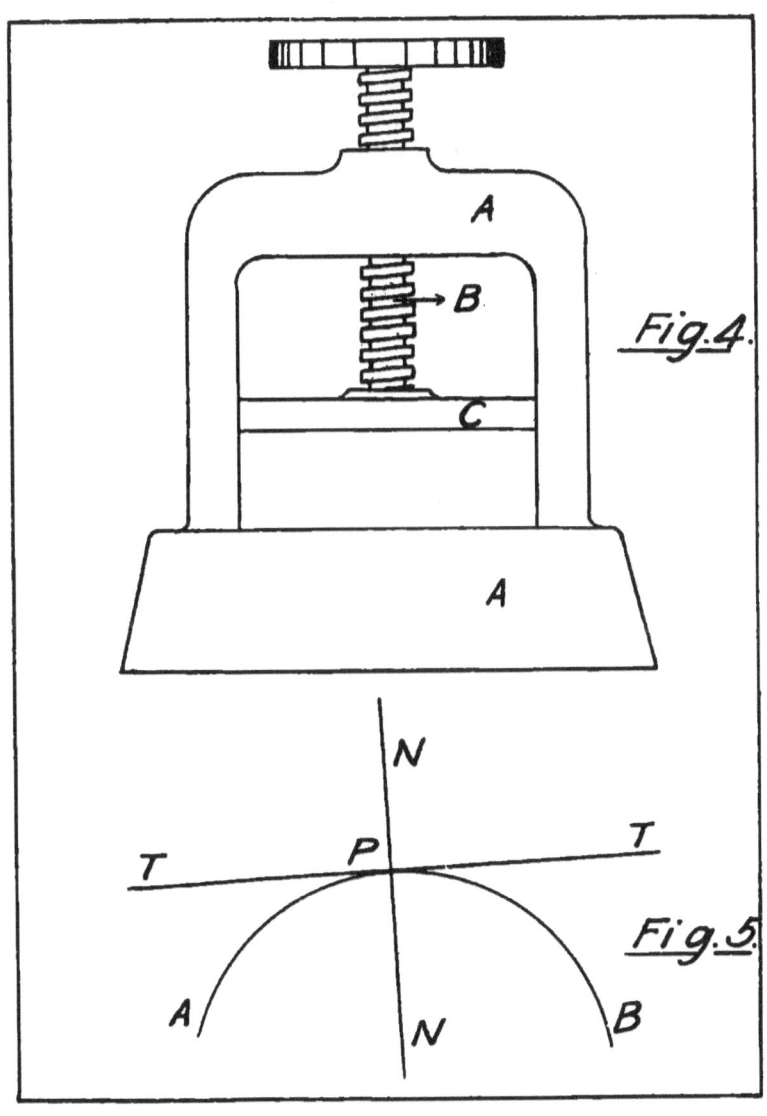

PRELIMINARY. 9

axis of the body's motion, and O is the instantaneous centre of the motion of the section shown in Fig. 6.

For the sake of brevity an instantaneous centre will be called a **centro**.

If TT and T_1T_1 had been parallel to each other, AB and CD would also have been parallel, and would have intersected at infinity; in which case the body's instantaneous motion would have been rotation about an axis infinitely distant; *i. e.*, it would have been *translation*.

The motion of the body in Fig. 6, is of course referred to a fixed body, which, in this case, may be represented by the paper. The instantaneous motion of the body is rotation about O relatively to the paper. Let M represent the figure, and N the fixed body represented by the paper. Suppose the material of M to be extended so as to include O. Then a pin could be put through O, materially connecting M and N, without interfering with their instantaneous motion. Such connection at any other point would interfere with the instantaneous motion.

The centro of the relative motion of two bodies is a point, and the only one, at which they have no relative motion; it is a point, and the only one, that is common to the two bodies for the instant.

It will be seen that the points of the figure in Fig. 6 might be moving in any paths, so long as those paths are tangent at the points to the lines representing the instantaneous motion.

In general, centros of the relative motion of two bodies are continually changing their position. They may, however, remain stationary; *i. e.*, they may become fixed centres of rotation.

8. Loci of Centros, or Centroids. — As centros change position they describe curves of some kind, and these loci of centros may be called **centroids**.

Suppose a section of any body, M, to have motion relatively to a section of another body, N (fixed), in the same or parallel plane. Centros may be found for a series of positions, and a curve drawn through them on the plane of N would be the centroid of the motion of M relatively to N. If, now, M being fixed, N moves so that the relative motion is the same as before, the centroid of the motion of N relatively to M, may be located upon the plane of M. Now, since the centro of the relative motion of two bodies is a point at which they have no relative motion, and since the points of the centroids become successively the centros of the relative motion, it follows that as the motion goes on, the centroids would roll upon each other without slipping. Therefore, if the centroids are drawn, and rolled upon each other without slipping, the bodies M and N will have the same relative motion as before. From this it follows that the relative plane motion of two bodies may be reproduced by rolling together, without slipping, the centroids of that motion.

9. Pairs of Motion Elements.—The external and internal surfaces by which motion is constrained in Figs. 2 and 3 may be called **pairs of motion elements.** The pair in Fig. 2 is called a **turning pair**, and the pair in Fig. 3 is called a **sliding pair.**

The helical surfaces by which a nut and screw engage with each other, are called a **twisting pair.** These three pairs of motion elements have their surfaces in contact throughout. They are called **lower pairs.** Another class, called **higher pairs,** have contact only along *elements* of their surfaces. *Examples.*—Cams and toothed wheels.

CHAPTER II.

MOTION IN MECHANISMS.

10. Linkages or Motion Chains; Mechanisms.
In Fig. 7, b is joined to c by a turning pair
    c "   " d " sliding   "
    d " a " turning   "
    a " b "   "   "
Evidently there is complete constrainment of the relative motion of a, b, c, and d. For, d being fixed, if any motion occurs in either a, b, or c, the other two must have a predetermined corresponding motion.

c may represent the cross-head, b the connecting-rod, and a the crank of a steam engine of the ordinary type. If c were rigidly attached to a piston upon which the expansive force of steam acts toward the right, a must rotate about ad. This represents a *machine*. The members a, b, c, and d, may be represented for the study of relative motions by the diagram, Fig. 8.

This assemblage of bodies, connected so that there is complete constrainment of motion, may be called a **motion chain** or **linkage**, and the connected bodies may be called **links**. The chain shown is a **simple chain**, because no link is joined to more than two others. If any of the links of a chain are joined to more than two others, the chain is a **compound chain**. Examples will be given later.

When one link of a chain is fixed, *i.e.*, when it becomes the standard to which the motion of the others is referred, the chain is called a **mechanism**. Fixing different links of a chain gives different mechanisms. Thus in Fig. 8, if d be fixed, the mechanism is that which is used in the usual type of steam engine, as in Fig. 7. It is called the **slider crank mechanism**.

But if a be fixed, the result is an entirely different mechanism; for b would then rotate about the permanent centre ab, d would rotate about the permanent centre ad, while c would have a more complex motion, rotating about a constantly changing centro, whose path may be found.

Fixing b or c would give, in each case, a different mechanism.

11. Location of Centros.—In Fig. 8 d is fixed and it is required to find the centros of rotation, either permanent or instantaneous, of the other three links. The motion of a, relatively to the fixed link d, is rotation about the fixed centre ad. The motion of c relatively to d is translation, or rotation about a centro cd, at infinity vertically. The link b has a point in common with a; it is the centro, ab, of their relative motion. This point may be considered as a point in a or b; in either case it can have but one direction of motion. As a point in a its motion, relatively to d, is rotation about ad. For the instant, then, it is moving along a tangent to the circle through ab. But as a point in b, its direction of instantaneous motion must be the same, and hence its motion must be about some point in the line $ad-ab$, extended if necessary. Also b has a point, bc, in common with c; and by the same reasoning as above, bc, as a point in b, rotates, for the instant, about some point of the vertical line through bc. Now ab and bc are points of a rigid body, and one rotates for the instant about some point of AB; and the other rotates for the instant about some point, CD; hence both (as well as all other points of b) must rotate about the intersection of AB and CD. Hence bd is the centro of the motion of b relatively to d.

The motion of a may be referred to c (fixed), and ac will be found (by reasoning like that applied to b) to lie at the intersection of the lines EF and GH.

The motion chain in Fig. 8, as before stated, is called the **slider crank chain**.

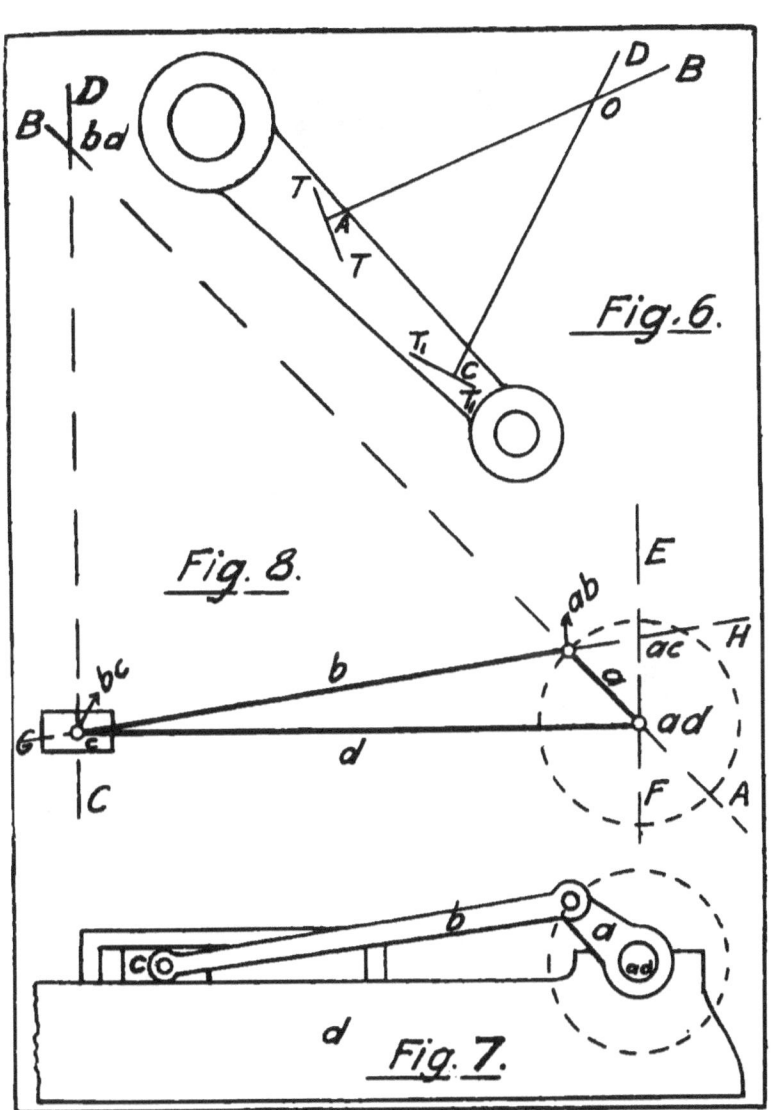

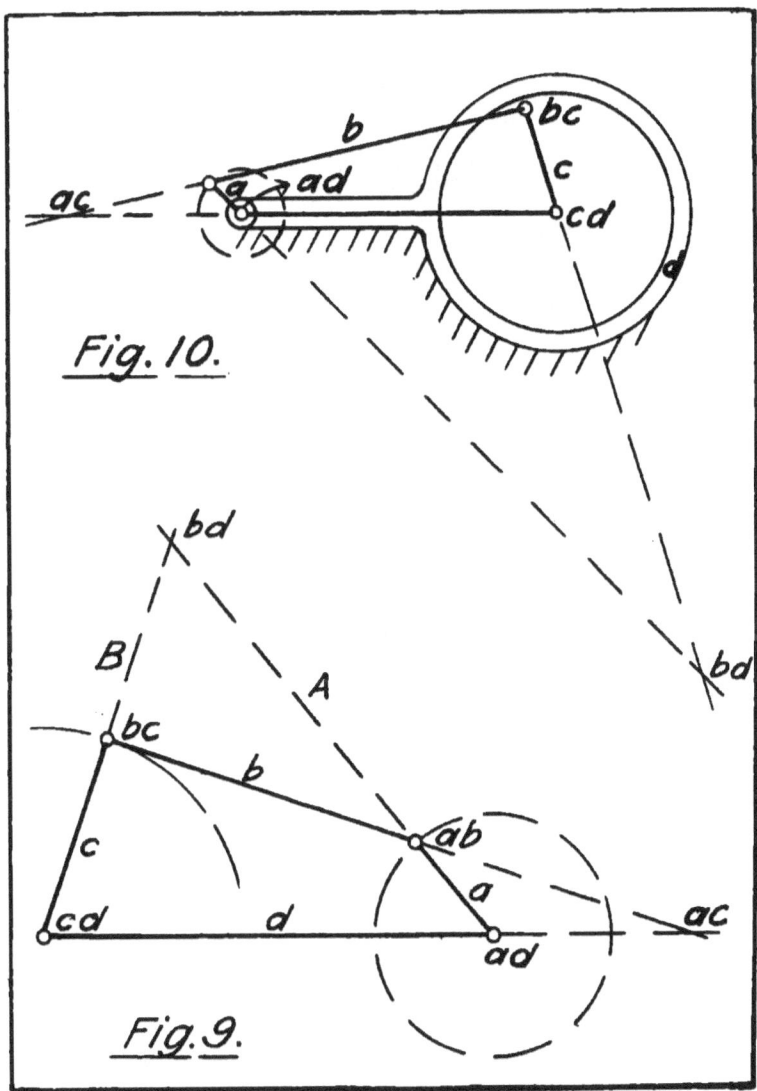

12. Centros of the Relative Motion of Three Bodies Are Always in the Same Straight Line. — In Fig. 8 it will be seen that the three centros of any three links lie in the same straight line. Thus ad, ab, and bd, are the centros of the links a, b, and d. This is true of any other set of three links. *Proof.*— Consider $a, b,$ and c. The centro ab as a point in a has a direction of instantaneous motion perpendicular to a line joining it to ad. As a point in b it has a direction of instantaneous motion perpendicular to a line joining it to bd. Therefore the lines $ab-ad$ and $ab-bd$ are both perpendicular to the direction of instantaneous motion of ab, and they also both pass through ab; hence they must coincide, and therefore ab, ad, and bd must lie in the same straight line. But $a, b,$ and d might be any three bodies whatever, which have relative plane motion, and the above reasoning would hold. Hence it may be stated: *The three centros of any three bodies having relative plane motion, must lie in the same straight line.* [The statement and proof of this important proposition is due to Prof. Kennedy.]

13. Lever Crank Chain. Location of Centros. — Fig. 9 shows a chain of four links of unequal length joined to each other by turning pairs. The centros ab, ad, cd, and bc may be located at once, since they are the centros of turning pairs which join adjacent links to each other. The centros of the relative motion of b, c, and d are bc, cd, and bd, and these must be in the same straight line. Hence bd is in the line B. The centros of the relative motion of a, b, and d, are ab, bd, and ad; and these also must lie in the same straight line. Hence bd is in the line A. Being at the same time in A and B, it must be at their intersection.

14. The Constrainment of Motion in a linkage is **independent of the size of the motion elements.** As long as the cylindrical surfaces of turning pairs have their axes unchanged, the surfaces themselves may be of any size whatever, and the motion is unchanged. The same is true of sliding and twisting pairs.

In Fig. 10, suppose the turning pair connecting c and d to be enlarged so that it includes bc. The link c now becomes a cylinder, turning in a ring attached to, and forming part of, the link b. bc

becomes a pin made fast in c and engaging with an eye at the end of b. The centros are the same as before the enlargement of cd, and hence the relative motion is the same.

In Fig. 11, the circular portion immediately surrounding cd is attached to d. The link c now becomes a ring moving in a circular slot. This may be simplified as in Fig. 12, whence c becomes a curved block moving in a limited circular slot in d. The centros remain as before, the relative motion is the same, and the linkage is essentially unchanged.

15. If, in the slider crank mechanism, the turning pair whose axis is ab, be enlarged till ad is included, as in Fig. 13, the motion of the mechanism is unchanged, but the link a is now called an eccentric instead of a crank. This mechanism is usually used to communicate motion from the main shaft of a steam engine to the valve. It is used because it may be put on the main shaft anywhere, without interfering with its continuity and strength.

The mechanism shown in Fig. 14 is called the "*slotted crosshead mechanism.*" Its centros may be found from principles already given.

This mechanism is often used as follows : One end of c, as E, is attached to a piston working in a cylinder attached to d. This piston is caused to reciprocate by the expansive force of steam or some other fluid. The other end of c is attached to another piston, which also works in a cylinder attached to d. This piston may pump water, or compress gas. The crank a is attached to a shaft, the projection of whose axis is ad. This shaft also carries a flywheel which insures approximately uniform rotation.

16. Location of Centros in a Compound Mechanism.—It is required to find the centros of the compound linkage, Fig. 15. In any linkage, each link has a centro relatively to every other link ; hence, if the number of links $= n$, the number of centros $= n(n-1)$. But the centro ab is the same as ba ; i. e., each centro is double. Hence the number of centros to be located for any linkage $= \dfrac{n(n-1)}{2}$. In the linkage Fig. 15, the number of centros $= \dfrac{6 \times 5}{2} = 15$.

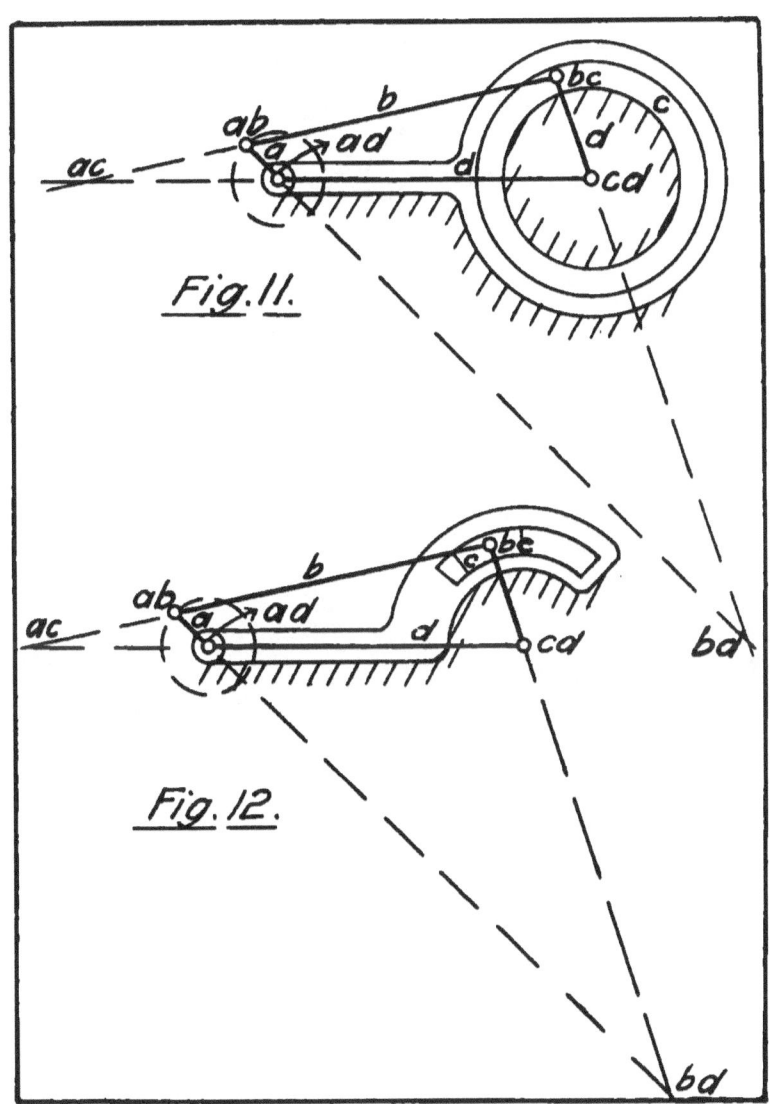

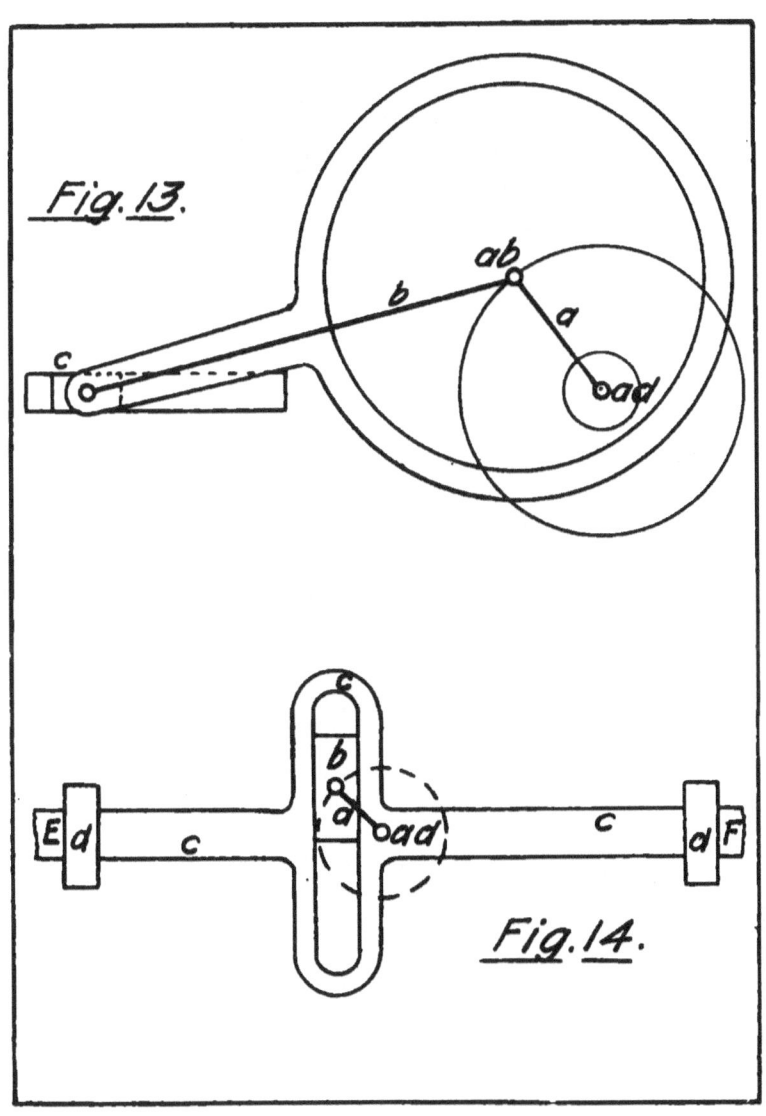

MOTION IN MECHANISMS. 15

The portion above the link d is a slider crank chain, and the character of its motion is in no way affected by the attachment of the part below d. On the other hand, the lower part is a lever crank chain, and the character of its motion is not affected by its attachment to the upper part. The chain may therefore be treated in two parts, and the centros of each part may be located from what has preceded. Each part will have six centros, and twelve would thus be located. ad, however, is common to the two parts, and hence only eleven are really found. Four centros, therefore, remain to be located. They are be, cf, bf, and ce. To locate be, consider the three links a, b, and e, and it follows that be is in the line A passing through ab and ae; considering b, d, and e, it follows that be is in the line B through bd and de. Hence be is at the intersection of A and B. Similar methods locate the other centros.

In general, for finding the centros of a compound linkage of six links, consider the linkage to be made up of two simple chains, and find their centros independently of each other. Then take the two links whose centro is required, together with one of the links carrying three motion elements (as a, Fig. 15). The centros of these links locate a straight line, A, which contains the required centro. Then take the two links whose centro is required, together with the other link which carries three motion elements. A straight line, B, is thereby located, which contains the required centro, and the latter is therefore at the intersection of A and B.

17. Velocity is rate of motion, or motion per unit time.

Linear velocity is linear space moved through in unit time; it may be expressed in any units of length and time; as miles per hour, feet per minnte or per second, etc.

Angular velocity is angular space moved through in unit time. In machines, angular velocity is usually expressed in revolutions per minute or per second.

The linear space described by a point in a rotating body, or its linear velocity, is directly proportional to its radius, or its distance from the axis of rotation. This is true because arcs are proportional to radii.

If A and B are two points in a rotating body, and if r_1 and r_2 are their radii, then the ratio of linear velocities

$$=\frac{\text{linear veloc. } A}{\text{linear veloc. } B} = \frac{r_1}{r_2}.$$

This is true whether the rotation is about a centre or a centro; *i. e.*, it is true both for continuous or instantaneous rotation. Hence it applies to all cases of plane motion in machines; because all plane motion in machines is equivalent to either continuous or instantaneous rotation about some point.

To find the relation of linear velocity of two points in a machine member, therefore, it is only necessary to find the relation of the radii of the points. The latter relation can easily be found when the centre or centro is located.

18. A vector is something which has magnitude and direction. A vector may be represented by a straight line, because the latter has magnitude (its length) and direction. Thus the length of a straight line, AB, may represent, upon some scale, the magnitude of some vector, and it may represent the vector's direction by being parallel to it, or by being perpendicular to it. For convenience the latter plan will here be used. The vectors to be represented are the linear velocities of points in mechanisms. The lines which represent vectors are also called vectors.

A line which represents the linear velocity of a point, will be called the linear velocity vector of the point. The symbol of linear velocity will be Vl. Thus VlA is the linear velocity of the point A. Also Va will be used as the symbol of angular velocity.

If the linear velocity and radius of a point are known, the angular velocity, or the number of revolutions per unit time, may be found; since the linear velocity $\div$ length of the circumference in which the point travels = angular velocity.

All points of a rigid body have the same angular velocity.

If the radii, and ratio of linear velocities of two points, in different machine members, are known, the ratio of the angular velocities of the members may be found as follows:

Let A be a point in a member M, and B a point in a member N.

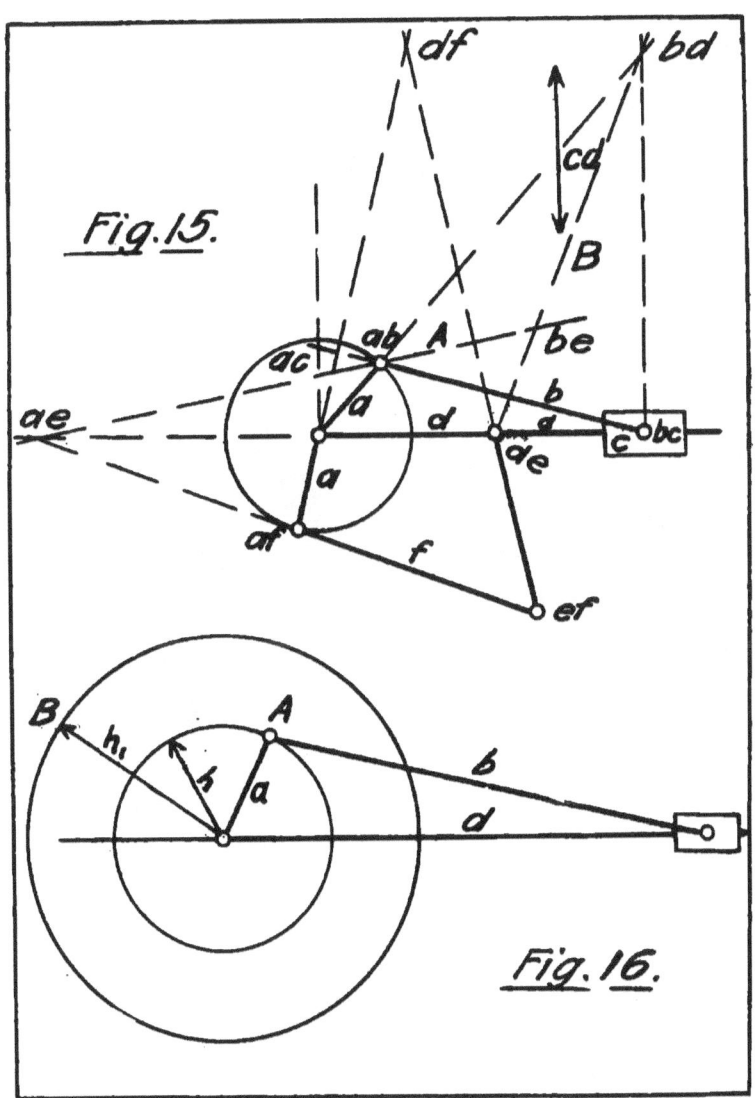

MOTION IN MECHANISMS.   17

$r_1 =$ radius of A; $r_2 =$ radius of B. VlA and VlB represent the linear velocities of A and B, whose ratio, $\dfrac{VlA}{VlB}$, is known.

Then $\qquad VaA = \dfrac{VlA}{2\pi r_1}$ and $VaB = \dfrac{VlB}{2\pi r_2}$.

Hence $\qquad \dfrac{VaA}{VaB} = \dfrac{VlA}{2\pi r_1} \times \dfrac{2\pi r_2}{VlB} = \dfrac{VlA}{VlB} \times \dfrac{r_2}{r_1} = \dfrac{VaM}{VaN}$.

If M and N rotate uniformly about fixed centres, the ratio $\dfrac{VaM}{VaN}$ is constant. If either M or N rotates about a centro, the ratio is a varying one.

19. To find the relation of linear velocity of two points in the same link, it is only necessary to measure the radii of the points, and the ratio of these radii is the ratio of the linear velocities of the points.

In Fig. 16, let the smaller circle represent the path of A, the centre of the crank pin of a slider crank mechanism; the link d being fixed. Let the larger circle represent the rim of a pulley, which is keyed to the same shaft as the crank. The pulley and the crank are then parts of the same link. The ratio of velocity of the crank pin centre and the pulley surface $= \dfrac{VlA}{VlB} = \dfrac{r}{r_1}$. In this case the link rotates about a fixed centre. The same relation holds, however, when the link rotates about a centro.

20. In Fig. 17, the link d is fixed and $\dfrac{Vl\,ab}{Vl\,bc} = \dfrac{ab - bd}{bc - bd}$.

By similar triangles this expression is also equal to $\dfrac{ab - O}{O - A}$.

Hence, if the radius of the crank circle be taken as the vector of the constant linear velocity of ab, the distance cut off on the vertical through O by the line of the connecting-rod (extended if necessary) will be the vector of the linear velocity of bc. Project A horizontally upon $bc - bd$, locating B. Then $bc - B$ is the vector of Vl of the slider, and may be used as an ordinate of the linear velocity diagram of the slider. By repeating the above construction for a series of positions, the ordinates representing the Vl of bc for different positions of the slider may be found. A smooth curve through the extremities of these ordinates is the velocity curve, from which the Vls for all points of the slider's stroke may be read. The scale of velocities, or the linear velocity represented by one inch of ordinate, equals the constant linear velocity of ab divided by $O-ab$ in inches.

21. It is required to find Vl of bc during a cycle of action of the mechanism shown in Fig. 18, d being fixed, and Vl of ab being constant. The two points, ab and bc, may both be considered in the link b. All points in b move about bd relatively to the fixed link.

Hence
$$\frac{Vl\,ab}{Vl\,bc} = \frac{ab-bd}{bc-bd}.$$

But a line, as MN, drawn parallel to b cuts off on the radii portions which are proportiontal to the radii themselves, and hence proportional to the Vls of the points. Hence

$$\frac{Vl\,ab}{Vl\,bc} = \frac{ab-M}{bc-N}.$$

The arc in which bc moves may be divided into any number of parts, and the corresponding positions of ab may be located. A circle through M, about ad, may be drawn, and the constant radial distance $ab-M$ may represent the constant velocity of ab. Through M_1, M_2, etc., draw lines parallel to the corresponding positions of b, and these lines will cut off on the corresponding line of c a distance which represents Vl of bc. Through the points thus determined the velocity diagram may be drawn, and the Vl of bc for a complete cycle is determined. The scale of velocities is found as in § 20.

22. The relation of linear velocity of points not in the same link may also be found.

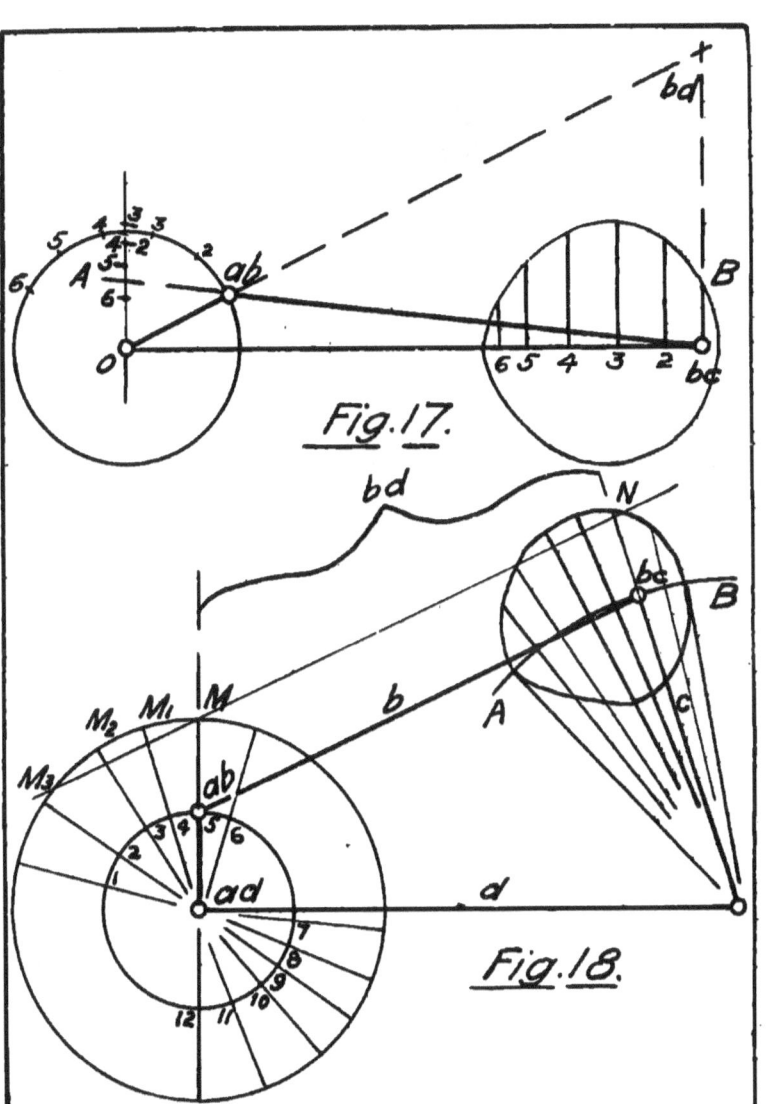

Fig.17.

Fig.18.

MOTION IN MECHANISMS.   19

Required $\dfrac{Vl \text{ of } A}{Vl \text{ of } B}$. The centro ab is a point common to a and b, the two links considered; d is the fixed link. Consider ab as a point in a; and its Vl is to that of A as their radii or distances from ad. Draw a vector triangle with its sides parallel to the triangle formed by joining A, ab, and ad. Then if the side A_1 represent the Vl of A, the side ab_1 will represent the Vl of ab. Consider ab as a point in b, and its Vl is to that of B as their radii, or distances to bd. Upon the vector ab_1, draw a triangle whose sides are parallel to those of a triangle formed by joining ab, bd, and B. Then, from similar triangles, the side B_1 is the vector of B's linear velocity.

Hence $\dfrac{Vl \text{ of } A}{Vl \text{ of } B} = \dfrac{\text{vector } A_1}{\text{vector } B_1}$.

The path of B during a complete cycle may be traced, and the Vl for a series of points may be found, by the above method; then the vectors may be laid off on normals to the path through the points; the velocity curve may be drawn; and the velocity of B at all points becomes known.

23. The diagram of Vl of the slider of the slider crank mechanism, Fig. 17, is unsymmetrical with respect to a vertical axis through its centre. This is due to the angularity of the connecting-rod, and may be explained as follows:

In Fig. 20, A is one angular position of the crank, and B is the corresponding angular position on the other side of the vertical through the centre of rotation. The corresponding positions of the slider are as shown. But for position A, the line of the connecting-rod, C, cuts off on the vertical through O, a vector Oa, which represents the slider's velocity. For position B the vector of the slider's velocity is Ob. Obviously this difference is due to the angularity of the connecting-rod.

In a mechanism which is equivalent to the slider crank with the connecting-rod always horizontal (as the slotted cross-head) the line of the connecting-rod would cut off on OY the same vector for position A and position B. Hence the velocity diagram for the slotted cross-head mechanism is symmetrical with respect to both vertical and horizontal axes through its centre. In fact, if the crank radius (= length of link a) be taken as the vector of the Vl of ab, the linear velocity diagram of the slider becomes a circle whose radius = the length of the link a. Hence the crank circle itself serves for the linear velocity diagram, the horizontal diameter representing the path of the slider.

24. During a portion of the cycle of the slider crank mechanism, the slider's Vl is greater than that of ab. This is also due to the angularity of the connecting-rod, and may be explained as follows : In Fig. 21, as the crank moves up from the position x, it will reach such a position, A, that the line of the connecting-rod extended will pass through B. OB in this position is the vector of the linear velocity of both ab and the slider, and hence their linear velocities are equal. When ab reaches B, the line of the connecting-rod passes through B; and again the vectors — and hence the linear velocities — of ab and the slider are equal. For all positions between A and B, the line of the connecting-rod will cut OB outside of the crank circle; and hence the linear velocity of the slider will be greater than that of ab. Obviously, the linear velocity of the slider is greatest when the angle between crank and connecting-rod $= 90°$. This result is due to the angularity of the connecting-rod, because if the latter remained always horizontal, its line could never cut OB outside the circle. It follows that in the slotted cross-head mechanism the maximum Vl of the slider = the constant Vl of ab. The angular space BOA, Fig. 21, throughout which Vl of the slider is greater than the Vl of ab, increases with increase of angularity of the connecting-rod ; $i.\,e.$, it increases with the ratio

$$\frac{\text{length of crank}}{\text{length of connecting-rod}}.$$

25. A slider in a mechanism often carries a cutting tool, which cuts during its motion in one direction, and is idle during the return

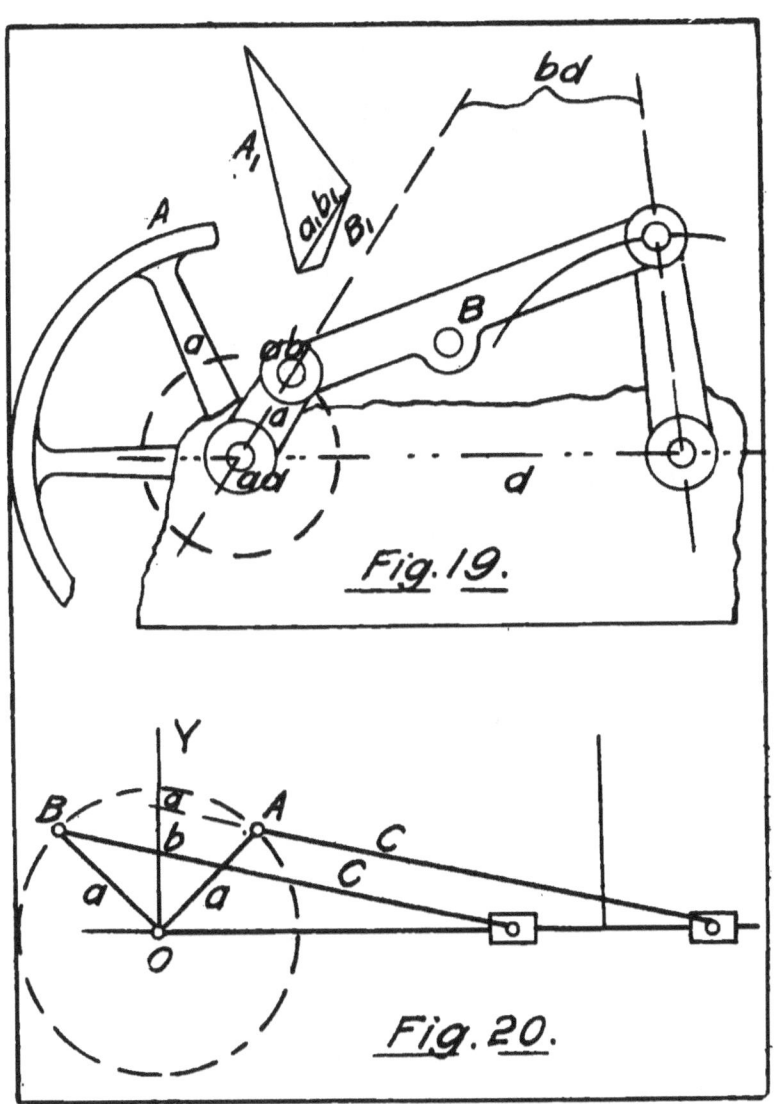

Fig. 19.

Fig. 20.

stroke. Sometimes the slider carries the piece to be cut, and the cutting occurs while it passes under a tool made fast to the fixed link, the return stroke being idle.

The velocity of cutting is limited. If the limiting velocity be exceeded, the tool becomes so hot that its temper is drawn, and it becomes unfit for cutting. The limit of cutting velocity depends on the nature of the material to be cut. Thus annealed tool-steel, and the scale surface of cast iron, may be cut at 20 feet per minute; wrought iron and soft steel at 25 to 30 feet per minute; while brass and the softer alloys may be cut at 40 or more feet per minute. There is no limit of this kind, however, to the velocity during the idle stroke; and it is desirable to make it as great as possible, in order to increase the product of the machine. This leads to the design and use of "quick return" mechanisms.

26. If, in a slider crank mechanism, the centre of rotation of the crank be moved, so that the line of the slider's motion does not pass through it, the slider will have a quick return motion.

In Fig. 22, when the slider is in its extreme position at the right, the crank-pin centre is at D. When the slider is at B, the crank-pin centre is at C. If rotation is as indicated by the arrow, then while the slider moves from B to A, the crank-pin centre moves from C over to D. And while the slider returns from A to B, the crank-pin centre moves under from D to C. If the Vl of the crank-pin centre be assumed constant, the time occupied in moving from D to C is less than that from C to D. Hence, the time occupied by the slider in moving from B to A is greater than that occupied in moving from A to B. The mean velocity during the forward stroke is therefore less than during the return stroke. Or the slider has a "quick return" motion.

It is required to design a mechanism of this kind for a length of stroke $= BA$ and for a ratio

$$\frac{\text{mean } Vl \text{ forward stroke}}{\text{mean } Vl \text{ return stroke}} = \frac{5}{7}.$$

The mean velocity of either stroke is inversely proportional to the time occupied, and the time is proportional to the corresponding angle described by the crank. Hence

$$\frac{\text{mean velocity forward}}{\text{mean velocity return}} = \frac{5}{7} = \frac{\text{angle } \beta}{\text{angle } \alpha}.$$

It is therefore necessary to divide 360° into two parts which are to each other as 5 to 7. Hence $\alpha = 210°$ and $\beta = 150°$. Obviously $\theta = 180° - \beta = 30°$. Place the 30° angle of a drawing triangle so that its sides pass through B and A. This condition may be fulfilled and yet the vertex of the triangle may occupy an indefinite number of positions. By trial O may be located so that the crank shall not interfere with the line of the slider. O being located tentatively, it is necessary to find the corresponding lengths of crank a, and connecting-rod b. When the crank-pin centre is at D, $AO = b - a$; when it is at C, $BO = b + a$. AO and BO are measurable values of length; hence a and b may be found, the crank circle may be drawn, and the velocity diagrams may be constructed as in Fig. 17; remembering that the distance cut off upon a vertical through O, by the line of the connecting-rod, is the vector of the Vl of the slider for the corresponding position, when the Vl of the crank-pin centre is represented by the crank radius.

The construction may be checked by finding the mean heights of the velocity diagram, or the areas, which are proportional to the mean heights, above and below the horizontal line; these should be to each other as 5:7. The areas may be found by use of a planimeter; and these areas, divided by the length of stroke, equal the mean heights.

It is required to make the maximum velocity of the forward stroke of the slider = 20 feet per minute, and to find the corresponding number of revolutions per minute of the crank. The maximum linear velocity vector of the forward stroke = the maximum height of the upper part of the velocity diagram; call it V_1. Call the linear velocity vector of the crank-pin centre V_2 = crank radius. Let x = linear velocity of the crank-pin centre. Then

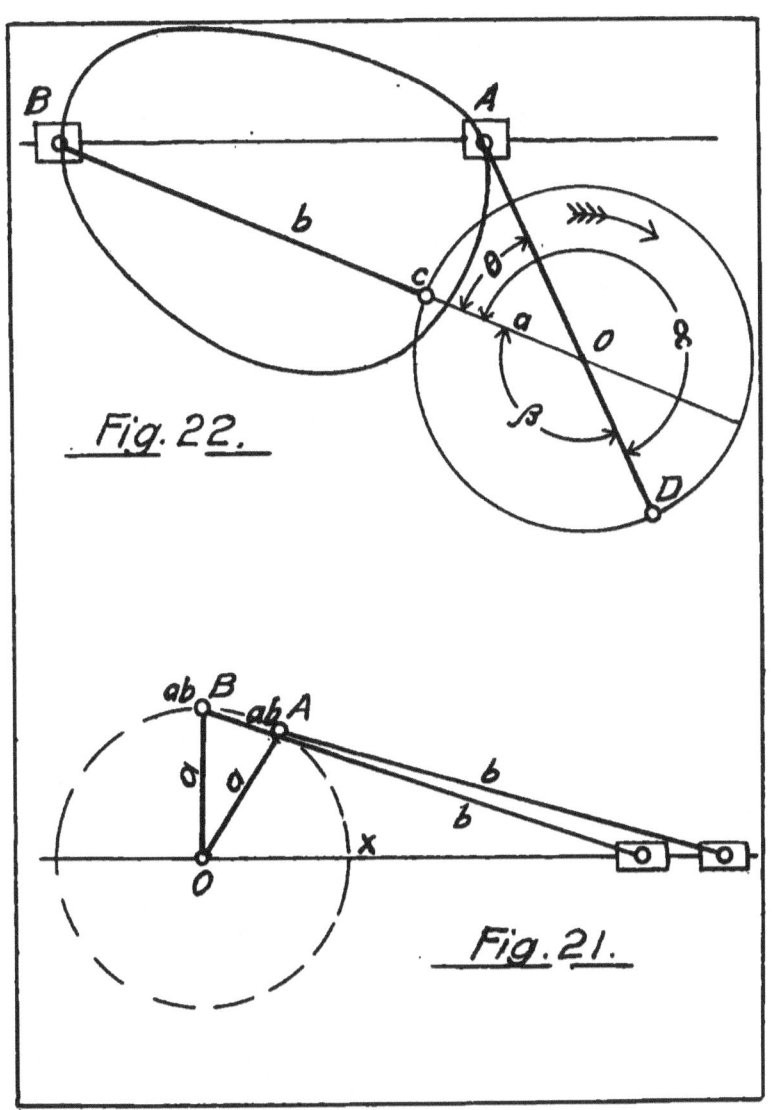

Fig. 22.

Fig. 21.

MOTION IN MECHANISMS.

$$\frac{V'_1}{V'_2} = \frac{20 \text{ ft. per minute}}{x},$$

or
$$x = \frac{20 \text{ ft. per minute} \times V'_2}{V'_1}.$$

x is therefore expressed in known terms. If now x, the space the crank-pin centre is required to move through per minute, be divided by the space moved through per revolution, the result will equal the number of revolutions per minute $= N$;

$$N = \frac{x}{2\pi \times \text{length of crank}}.$$

27. Fig. 23 shows a compound mechanism. The link d is the supporting frame, and a rotates about ad in the direction indicated, communicating motion to c through the slider b, so that c vibrates about cd. The link e, connected to c by a turning pair at ce, causes f to slide horizontally on another part of the frame or fixed link d. The centre of the crank-pin, ab, is given a constant linear velocity, and the slider, f, has motion toward the left with a certain mean velocity, and returns toward the right with a greater mean velocity. This is true because the slider f moves toward the left, while a moves through the angle α; and toward the right while a moves through the angle β. But the motion of a is uniform, and hence the angular movement α represents more time than the angular movement β; and f, therefore, has more time to move toward the left, than it has to move through the same space toward the right. It therefore has a "quick return" motion.

The machine is driven so that the crank-pin centre moves uniformly, and the velocity, at all points of its stroke, of the slider carrying a cutting tool, is required. The problem, therefore, is to find the relation of linear velocities of ef and ab, for a series of positions during the cycle; and to draw the diagram of velocity of ef.

Solution.—ab has a constant known linear velocity. The point in the link c which coincides, for the instant, with ab, receives motion from ab; but the direction of its motion is different from that of *ab*, because *ab* rotates about *ad* while the coinciding point of *c* rotates about *cd*. If *ab-A* be laid off representing the linear velocity of *ab*, then *ab-B* will represent the linear velocity of the coinciding point of the link *c*. Let the latter point be called *x*.

Locate *cf*, at the intersection of *e* with the line *cd — ad*. Now *cf* and *x* are both points in the link *c*, and hence their linear velocities, relatively to the fixed link, are proportional to their distances from *cd*. These two distances may be measured directly, and with the known value of linear velocity of $x = ab\text{-}B$, give three known values of a simple proportion, from which the fourth term, the linear velocity of *cf*, may be found.

Or, if the line *BD* be drawn parallel to *cd-ad*, the triangle *B-D-ab* is similar to the triangle *cd-cf-ab*, and from the similarity of these triangles, it follows that *BD* represents the linear velocity of *cf* on the same scale that *ab-B* represents the linear velocity of *x*. Hence the linear velocity of *cf*, for the assumed position of the mechanism, becomes known. But since *cf* is a point of the slider, all of whose points have the same linear velocity, it follows that the linear velocity of *cf* is the required linear velocity of the slider.

This solution may be made for as many positions of the mechanism as are necessary to locate accurately the velocity curve.

Having drawn the velocity diagram, suppose that it is required to make the maximum linear velocity of the slider on the slow stroke $= A$ feet per minute. Then the linear velocity of the crank-pin centre $ab = y = A \dfrac{\text{max. } Vl \text{ of slider}}{Vl \text{ of } ab}$. If $r =$ the crank radius, the number of revolutions per minute $= \dfrac{y}{2\pi r}$.

When this mechanism is embodied in a machine, *a* becomes a crank attached to a shaft whose axis is at *ad*. The shaft turns in bearings provided in the machine frame. The crank carries a pin whose axis is at *ab*, and this pin turns in a bearing in the sliding block *b*. The link *c* becomes a lever keyed to a shaft whose axis is at *cd*. This lever has a long slot in which the block *b* slides. The link e becomes a connecting-rod, connected both to c and f by pin and bearing. The link f becomes the "cutter bar" or "ram" of a shaper; the part which carries the cutting tool. The link d becomes the frame of the machine, which not only affords support to the shafts at ad and cd, and the guiding surfaces for f, but also is so designed as to afford means for holding the pieces to be planed, and supports the feeding mechanism.

28. Fig. 24 shows another compound linkage. d is fixed, and c rotates uniformly about cd, communicating rotary motion to a through the slider b. a is extended past ad (the part extended being in another parallel plane), and moves a slider f through a link e. This is called the "Whitworth quick return mechanism." The point bc at which c communicates motion to a moves along a, and hence the radius of the point at which a receives a constant linear velocity varies, and the angular velocity of a must vary inversely. Hence the angular velocity of a is a maximum when the radius is a minimum, $i.\,e.$, when a and c are vertical downward; and the angular velocity of a is minimum when the radius is a maximum, $i.\,e.$, when a and c are vertical upward.

29. *Problem.*—To design a Whitworth Quick Return for a given

ratio, $\quad \dfrac{\text{mean } Vl \text{ of } f \text{ forward}}{\text{mean } Vl \text{ of } f \text{ returning}}$.

When the centre of the crank-pin, C, reaches A, the point D will coincide with B, the link C will occupy the angular position $cd\text{-}B$, and the slider f will be at its extreme position toward the left.

When the point C reaches F, the point D will coincide with E, the link c will occupy the angular position $cd\text{-}E$, and the slider f will be at its extreme position toward the right.

Obviously, while the link c moves *over* from the position $cd\text{-}E$ to the position $cd\text{-}B$, the slider f will complete its forward stroke. While c moves *under* from $cd\text{-}B$ to $cd\text{-}E$, f will complete the return stroke. The link c moves with a uniform angular velocity, and hence the mean velocity of f forward is inversely proportional to the angle β, and the mean velocity of f returning is inversely proportional to α. Or $\dfrac{\text{mean } Vl \text{ of } f \text{ forward}}{\text{mean } Vl \text{ of } f \text{ returning}} = \dfrac{\alpha}{\beta}$.

For the design, the distance cd-ad must be known. This may usually be decided on from the limiting sizes of the journals at cd and ad. Suppose that the above ratio $= \dfrac{\alpha}{\beta} = \dfrac{5}{7}$, that cd-$ad = 3''$, and that the maximum length of stroke of $f = 12''$. Locate cd and measure off vertically downward a distance equal to 3'', thus locating ad. Draw a horizontal line through ad. The point ef of the slider f will move along this line.

Since $\dfrac{\alpha}{\beta} = \dfrac{5}{7}$, and $\alpha + \beta = 360°$,

$\therefore \alpha = 150°$ and $\beta = 210°$.

Lay off α from cd as a centre, so that the vertical line through cd bisects it. Draw a circle through B with cd as a centre, B being the point of intersection of the bounding line of α with a horizontal through ad. The length of the link $c = cd$-B.

The radius ad-C must equal the travel of $f \div 2 = 6''$. This radius is made adjustable, so that the length of stroke may be varied. The connecting-rod, e, may be made of any convenient length.

30. Problem. — To draw the velocity diagram of the slider f of the Whitworth Quick Return. The point bc, Fig. 25, has a known constant linear velocity, and its direction of motion is always at right angles to a line joining it to cd. The point of the link a, which coincides in this position of the mechanism with bc, receives motion from bc, but its direction of motion is at right angles to the line bc-ad. If bc-A represent the linear velocity of bc, its projection upon bc-ad extended will represent the linear velocity of the point of a which coincides with bc. Call this point x. The centro af may be considered as a point in a, and its linear velocity relatively to d, when so considered, is proportional to its distance from ad.

Hence $\dfrac{Vl \text{ of } af}{Vl \text{ of } x} = \dfrac{ad\text{-}af}{ad\text{-}bc}$.

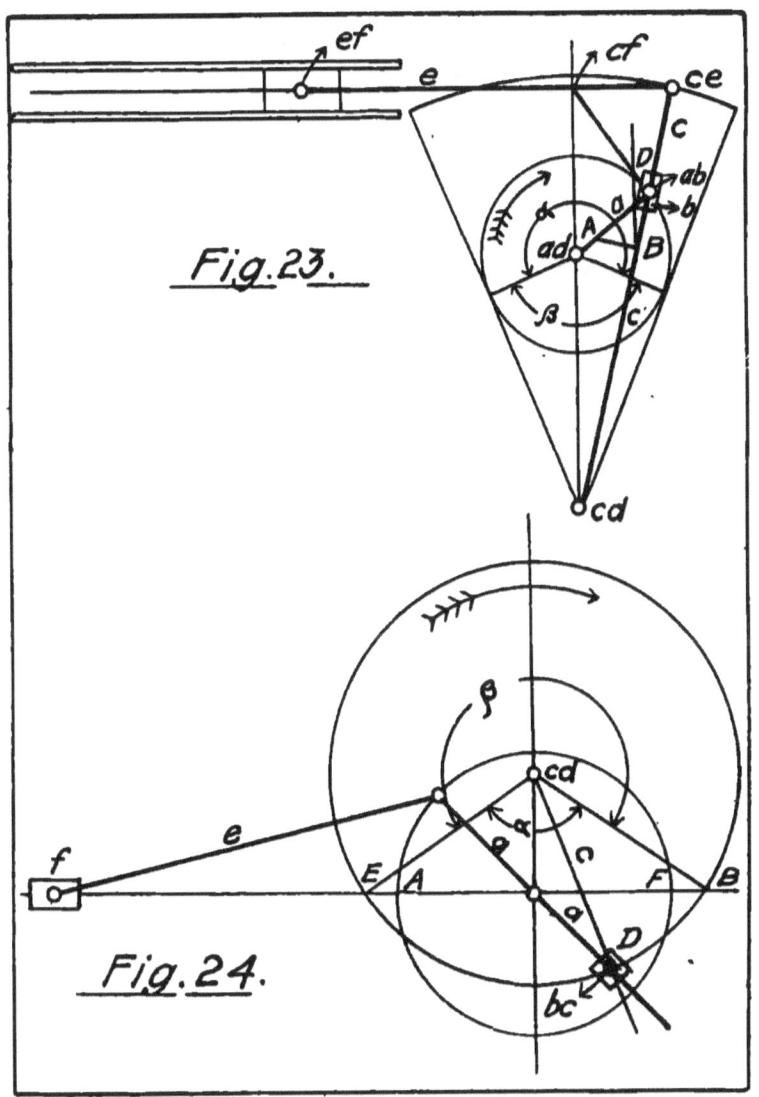

Fig. 23.

Fig. 24.

But the triangles *ad-cd-bc*, and *B-C-bc* are similar. Hence

$$\frac{Vl \text{ of } af}{Vl \text{ of } x} = \frac{BC}{B\text{-}bc}.$$

This means that BC represents the linear velocity of af upon the same scale that $B\text{-}bc$ represents the linear velocity of x. But af is a point in f, and all points in f have the same linear velocity; hence BC represents the linear velocity of the slider f, for the given position of the mechanism, and it may be laid off as an ordinate of the velocity curve. This solution may be made for as many positions as are required to locate accurately the entire velocity curve for a cycle of the mechanism.

CHAPTER III.

ENERGY IN MACHINES.

31. The subject of motion and velocity, in certain simple machines, has been treated and illustrated. It remains now to consider the passage of *energy* through similar machines. From this the solution of force problems will follow.

During the passage of energy through a machine, or chain of machines, any one, or all, of four changes may occur.

I. The energy may be transferred in space. *Example.* — Energy is received at one end of a shaft and transferred to the other end, where it is received and utilized by a machine.

II. The energy may be converted into another form. *Examples.* — (*a*) Heat energy into mechanical energy by the steam engine machine chain. (*b*) Mechanical energy into heat by friction. (*c*) Mechanical energy into electrical energy, as in a dynamo-electric machine; or electrical energy into mechanical energy in the electric motor, etc.

III. Energy is the product of a force factor and a space factor. Energy per unit time, or **rate of doing work**, is the product of a force factor and a velocity factor; since velocity is space per unit time. Either factor may be changed at the expense of the other; *i. e.*, velocity may be changed, if accompanied by such a change of force that the energy per unit time remains constant. Correspondingly, force may be changed at the expense of velocity, energy per unit time being constant. *Example.* — A belt transmits 6000 foot-pounds of energy per minute to a machine. The belt velocity is

120 feet per minute, and the force exerted is 50 lbs. Frictional resistance is neglected. A cutting tool in the machine does useful work; its velocity is 20 feet per minute, and the resistance to cutting is 300 lbs. Then, energy received per minute $= 120 \times 50 =$ 6000 foot-pounds; and energy delivered per minute $= 20 \times 300 =$ 6000 foot-pounds. The energy received therefore equals the energy delivered. But the velocity and force factors are quite different in the two cases.

IV. Energy may be transferred in time. In many machines the energy received at every instant equals that delivered. There are many cases, however, where there is a periodical demand for work, *i. e.*, a fluctuation in the rate of doing work; while energy can only be supplied at the average rate. Or there may be a uniform rate of doing work, and a fluctuating rate of supplying energy. In such cases means are provided in the machine, or chain of machines, for the *storing of energy* till it is needed. In other words, *energy is transferred in time*. *Examples.* — (*a*) In the steam engine there is a varying rate of supplying energy during each stroke, while there is (in general) a uniform rate of doing work. There is, therefore, a periodical excess and deficiency of effort. A heavy wheel on the main shaft absorbs the excess of energy with increased velocity, and gives it out again with reduced velocity, when the effort is deficient. (*b*) A pump delivers water into a pipe system under pressure. The water is used in a hydraulic press, whose action is periodic, and beyond the capacity of the pump. A hydraulic accumulator is attached to the pipe system, and while the press is idle the pump slowly raises the accumulator weight, thereby storing potential energy, which is given out rapidly by the descending weight for a short time while the press acts. (*c*) A dynamo-electric machine is run by a steam engine, and the electrical energy is delivered and stored in storage batteries, upon which there is a periodical demand. In this case, as well as in case (*b*), there is a transformation of energy as well as a transfer in time.

32. Force Problems.—Suppose the slider crank mechanism in Fig. 26 to represent a shaping machine; the velocity diagram of the slider being drawn. The resistance offered to cutting metal during the forward stroke must be overcome. This resistance may be assumed constant. Throughout the cutting stroke there is a constantly varying *rate of doing work*. This is because the rate of doing work = resisting force (constant) × velocity (varying). This product is constantly varying, and is a maximum when the slider's velocity is a maximum. The slider must be driven by means of energy transmitted through the crank a. The maximum rate at which energy must be supplied, equals the maximum rate of doing work at the slider. Draw the mechanism in the position of maximum velocity of slider; *i. e.*, locate the centre of the slider-pin at the base of the maximum ordinate of the velocity diagram, and draw b and a in their corresponding positions. The slider's known velocity is represented by y, and the crank-pin's required velocity is represented by z on the same scale. Hence the value of x becomes known by simple proportion. The rate of doing work must be the same at c and at ab (neglecting friction).* Hence $Rv_1 = Fv_2$, in which R and v_1 represent the force and velocity factors at c; and F and v_2 represent the force and velocity factors at ab. R and v_1 are known from the conditions of the problem, and v_2 is found as above. Hence, F may be found, $= \dfrac{Rv_1}{v_2} =$ force which, applied tangentially to the crank-pin centre, will overcome the maximum resistance of the machine. In all other positions of the cutting stroke the rate of doing work is less, and F would be less. But it is necessary to provide driving mechanism capable of overcoming the maximum resistance, when no fly-wheel is used. If now F be multiplied by the crank radius, the product equals the maximum torsional moment ($= M$) required to drive the machine. If the energy is received on some different radius, as in case of gear or belt transmission, the maximum driving force $= M \div$ the new radius. During the return stroke the cutting tool is idle, and it is only necessary to

* The effect of acceleration to redistribute energy is zero in this position, because the acceleration of the slider at maximum velocity is zero, and the angular acceleration of b can only produce pressure in the journal at ad.

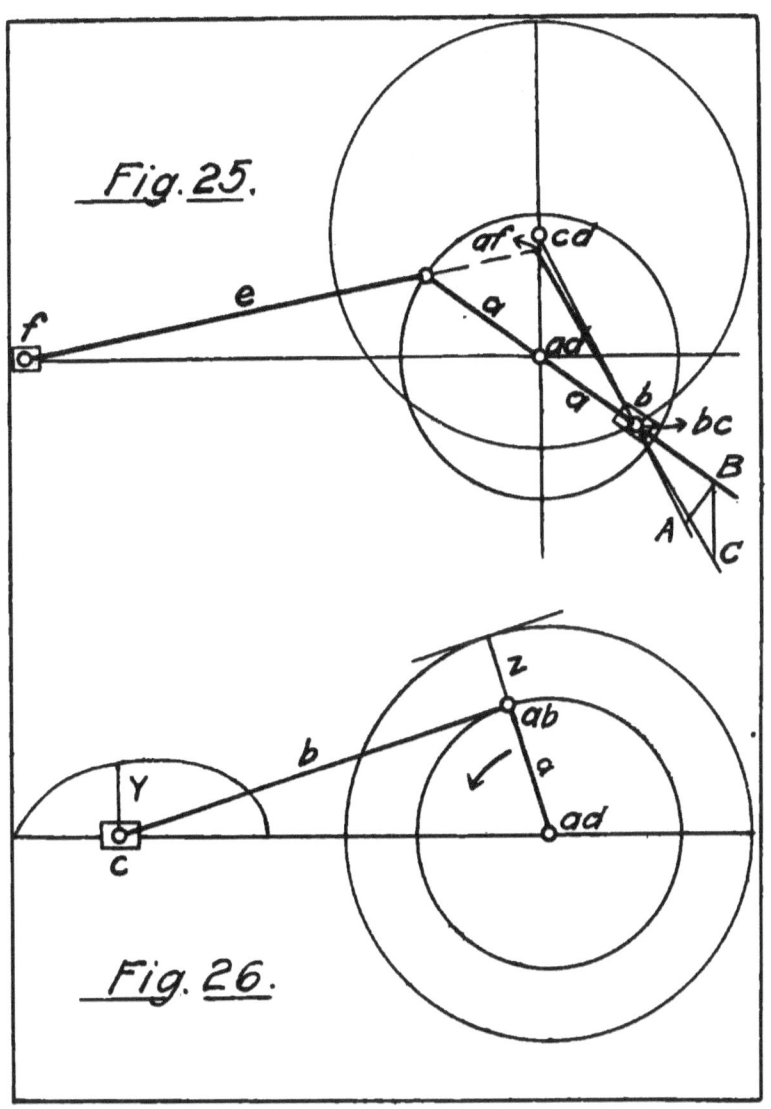

overcome the frictional resistance to motion of the bearing surfaces. Hence, the return stroke is not considered in designing the driving mechanism. When the *method* of driving this machine is decided on, the *capacity* of the driving mechanism must be such that it shall be capable of supplying to the crank shaft the torsional driving moment M, determined as above.

This method applies as well to the quick return mechanisms given. In each, when the velocity diagram is drawn, the vector of the maximum linear velocity of the slider, $= L_1$, and of the constant linear velocity of the crank-pin centre, $= L_2$, are known, and the velocities corresponding, V_1 and V_2, are also known, from the scale of velocities. The rate of doing work at the slider and at the crank-pin centre is the same, friction being neglected. Hence $Rv_1 = Fv_2$, or, since the vector lengths are proportional to the velocities they represent, $RL_1 = FL_2$; and $F = \dfrac{RL_1}{L_2}$. Therefore the resistance to the slider's motion $= R$, on the cutting stroke, multiplied by the ratio of linear velocity vectors, $\dfrac{L_1}{L_2}$, of slider and crank-pin, equals F, the maximum force that must be applied tangentially at the crank-pin centre to insure motion. F multiplied by the crank radius $=$ maximum torsional driving moment required by the crank shaft. If R is varying, and known, find where Rv, the rate of doing work, is a maximum, and solve for that position in the same way as above.

33. Force Problems — *Continued.* — In the usual type of steam engine the slider crank mechanism is used, but energy is supplied to the slider (which represents piston, piston-rod, and cross-head), and the resistance opposes the rotation of the crank and attached shaft. In any position of the mechanism, Fig. 28, force applied to the crank-pin through the connecting-rod, may be resolved into two components, one radial and one tangential. The tangential component tends to produce rotation; the radial component produces pressure between the surfaces of the shaft journal and its bearing. The tangential component is a maximum when the angle between crank and connecting-rod equals 90°; and it becomes zero when C reaches A or B. If there is a uniform resistance, the rate of doing work is constant. Hence, since the energy is supplied at a varying rate, it follows that during part of the revolution the effort is greater than the resistance; while during the remaining portion of the revolution the effort is less than the resistance. When effort is less than resistance, the machine will stop unless other means are provided to maintain motion. A "fly-wheel" is keyed to the shaft, and this wheel, because of slight variations of velocity, alternately stores and gives out the excess and deficiency of energy of the effort, thereby adapting it to the constant work to be done.

34. *Problem.* — Given length of stroke of the slider of a steam engine slider crank mechanism, the required horse-power, or rate of doing work, and number of revolutions. Required the total *mean* pressure that must be applied to the piston.

Let
L = length of stroke = 1 foot;
HP = horse-power = 20;
N = revolutions per minute = 200;
F = required mean force on piston.

Then $N \times L = 200$ feet per minute = mean velocity of slider = V. Now, the mean rate of doing work in the cylinder and at the main shaft during each stroke is the same (friction neglected); hence $FV = HP \times 33000$,

$$F = \frac{HP \times 33000}{V} = \frac{20 \times 33000}{200} = 3300 \text{ lbs.}$$

35. In the slider crank chain, the velocity of the slider necessarily varies from zero, at the ends of its stroke, to a maximum value near mid-stroke. The mass of the slider and attached parts is therefore positively and negatively accelerated each stroke. When a mass is positively accelerated it stores energy; and when it is negatively accelerated it gives out energy. The amount of this energy, stored or given out, depends upon the mass and the acceleration. The slider stores energy during the first part of its stroke, and gives it out during the second part of its stroke. While, there-

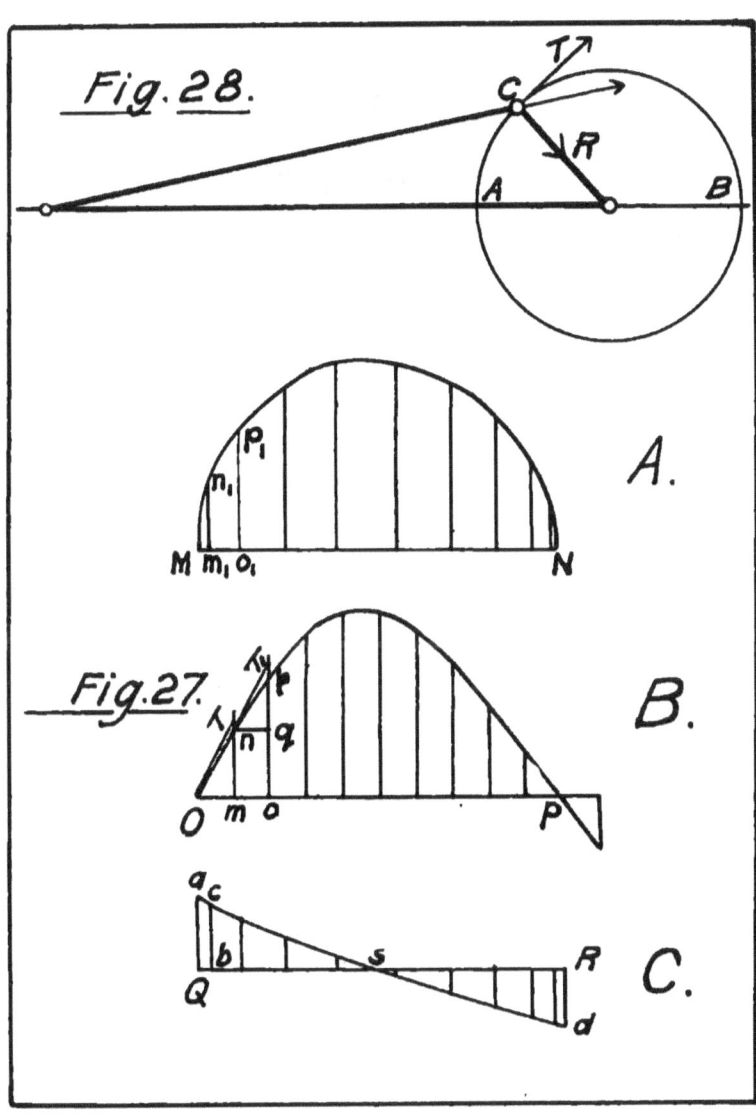

fore, it gives out all the energy it receives, it gives it out differently distributed. In order to find exactly how the energy is distributed, it is necessary to find the acceleration throughout the slider's stroke. This may be done as follows : Fig. 27 A shows the velocity diagram of the slider of a slider crank mechanism, for the forward stroke. The acceleration required at any point $\frac{\Delta v}{\Delta t}$, in which Δv is the increase in velocity during any interval of time Δt, assuming that the increase in velocity becomes constant at that point. Lay off the horizontal line OP MN. Divide OP into as many *equal* parts as there are *unequal* parts in MN. These divisions may each represent Δt. At m erect the ordinate $mn = m_1n_1$; and at o erect the ordinate op - o_1p_1. Continue this construction throughout OP, and draw a curve through the upper extremities of the ordinates. B is a velocity diagram on a "time base." At O draw the tangent OT to the curve. If the increase in velocity were constant during the time interval represented by Om, the increment of velocity would be represented by mT. Therefore mT is proportional to the acceleration at the point O, and may be laid off as an ordinate of an acceleration diagram C. Thus $Qa = mT$. The divisions of QR are the same as those of MN; i.e., they represent *positions* of the slider. This construction may be repeated for the other divisions of the curve B. Thus at n the tangent nT_1 and horizontal nq are drawn, and qT_1 is proportional to the acceleration at n, and is laid off as an ordinate bc of the acceleration diagram. To find the value in acceleration units of Qa, mT is read off in velocity units $= \Delta v$, by the scale of ordinates of the velocity diagram. This value is divided by Δt, the time increment corresponding to Om. The result of this division $\frac{\Delta v}{\Delta t}$ acceleration at M in acceleration units. $\Delta t =$ the time of one stroke, or of one-half revolution of the crank, divided by the number of divisions in OP. If the linear velocity of the centre of the crank-pin, $= V$, be represented by the length of the crank radius $\frac{MN}{2} = r$, then the scale of velocities, or velocity in feet per second for 1" of ordinate $=\dfrac{V}{r}=\dfrac{\pi DN}{r\,60}$. D is the actual diameter of the crank circle, and r is the crank radius measured on the figure. If the weight, W, of parts accelerated is known, the force, F, necessary to produce the acceleration at any slider position may be found from the fundamental formula of mechanics

$$F = Mp = \dfrac{Wp}{g}$$

p being the acceleration corresponding to the position considered. If the ordinates of the acceleration diagram be taken as representing the *forces* which produce the acceleration, the diagram will have force ordinates and space abscissæ, and areas will represent work. Thus, Qas represents the work stored during acceleration, and Rsd represents the work given out during retardation. Let MN, Fig. 29, represent the length of the slider's stroke, and NC the resistance of cutting (uniform); then energy to do cutting per stroke is represented by the area $MBCN$. But during the early part of the stroke the reciprocating parts must be accelerated, and the force necessary at the beginning, found as above, $= BD$. The driving gear must, therefore, be able to overcome resistance equal to $MB + BD$. The acceleration, and hence the accelerating force, decreases as the slider advances, becoming zero at E. From E on the acceleration becomes negative, and hence the slider gives out energy and helps to overcome the resistance, and the driving gear has only to furnish energy represented by the area $AEFN$, though the work really done against resistance equals that represented by the area $CEFN$. The energy represented by the difference of these areas, $= ACE$, is that which is stored in the slider's mass during acceleration. Since by the law of conservation of energy, energy given out per cycle $=$ that received, it follows that area $ACE =$ area DEB, and area $BCMN = ADMN$. This redistribution of energy would seem to modify the problem on page 30, since that problem is based on the assumption of uniform resistance during cutting stroke. The position of maximum velocity of slider, however, corresponds to acceleration $=0$. The maximum rate of doing work, and the corresponding torsional driving moment at the crank shaft would probably correspond to the same position, and would not be materially changed. In such machines as shapers, the acceleration and weight of slider are so small that the redistribution of energy is unimportant.

36. Solution of the force problem in the steam engine slider crank mechanism; slider represents piston with its rod, and the crosshead.— The steam acts upon the piston with a pressure which varies during the stroke. The pressure is redistributed before reaching the cross-head pin, because the reciprocating parts are accelerated in the first part of the stroke, with accompanying storing of energy and reduction of pressure on the cross-head pin; and retarded in the second part of the stroke, with accompanying giving out of energy and increase of pressure on the cross-head pin. Let the ordinates of the full line diagram above OX, Fig. 30 A, represent the effective pressure on the piston throughout a stroke. B is the velocity diagram of slider. Find the acceleration throughout stroke, and from this and the known value of weight of slider, find the force due to acceleration. Draw diagram C, whose ordinates represent the force due to acceleration, upon the same force scale used in A. Lay off this diagram on OX as a base line, thereby locating the dotted line. The vertical ordinates between this dotted line and the upper line of A represent the pressure applied to the cross-head pin. These ordinates may be laid off from a horizontal base line, giving D. The product of the values of the corresponding ordinates of B and $D =$ the *rate of doing work throughout the stroke*. Thus the value of GH in pounds $\times$ value of EF in feet per second $=$ the rate of doing work in foot-pounds per second upon the cross-head pin, when the centre of the cross-head pin is at E. The rate of doing work at the crank-pin is the same as at the cross-head pin. Hence dividing this rate of doing work, $= EF \times GH$, by the constant tangential velocity of the crank-pin centre, gives the force acting tangentially on the crank-pin to produce rotation. The tangential forces acting throughout a half revolution of the crank may be thus found, and plotted upon a horizontal base line = length of half the crank circle, Fig. 31. The work done upon the piston, cross-head pin, and crank during a piston stroke is the same. Hence the areas of A and D, Fig. 30, are equal to each other, and to the area of the diagram, Fig. 31. The forces acting along the connecting-rod for all positions during the piston stroke, may be found by drawing force triangles with one side horizontal, one vertical, and one parallel to position of connecting-rod axis, the horizontal side being equal to the corresponding ordinate of D. The vertical sides of these triangles will represent the guide reaction, while the side parallel to the connecting-rod axis represents the force transmitted by the connecting-rod.

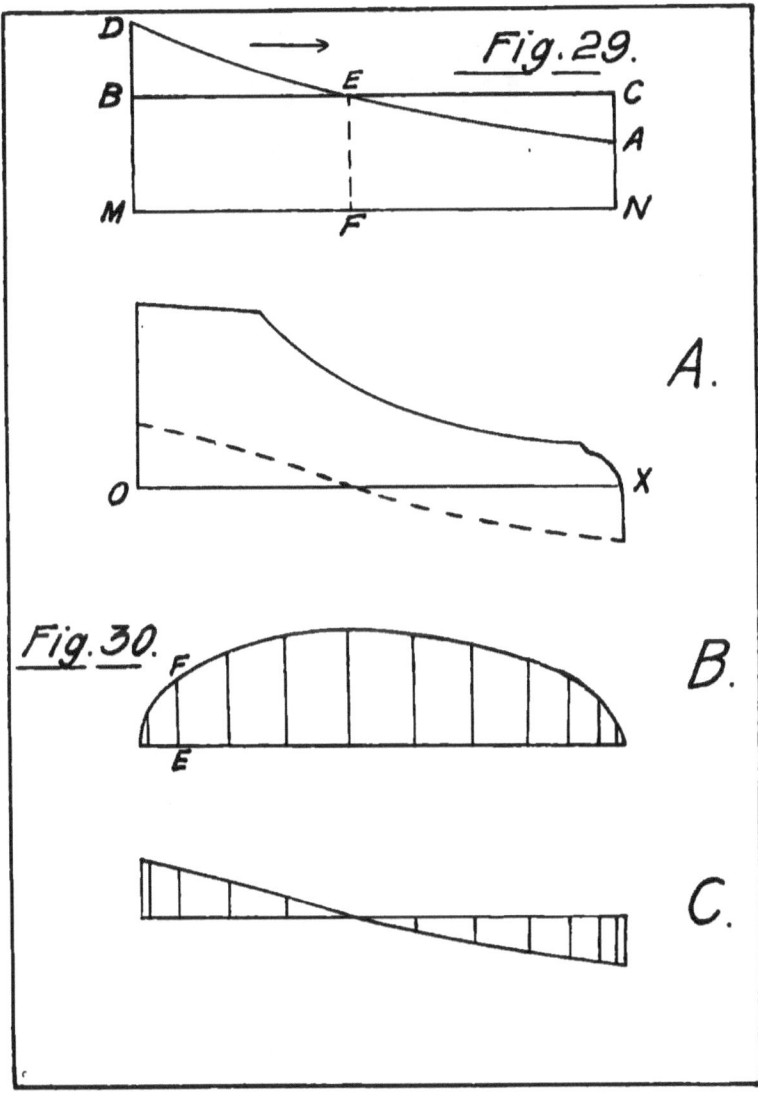

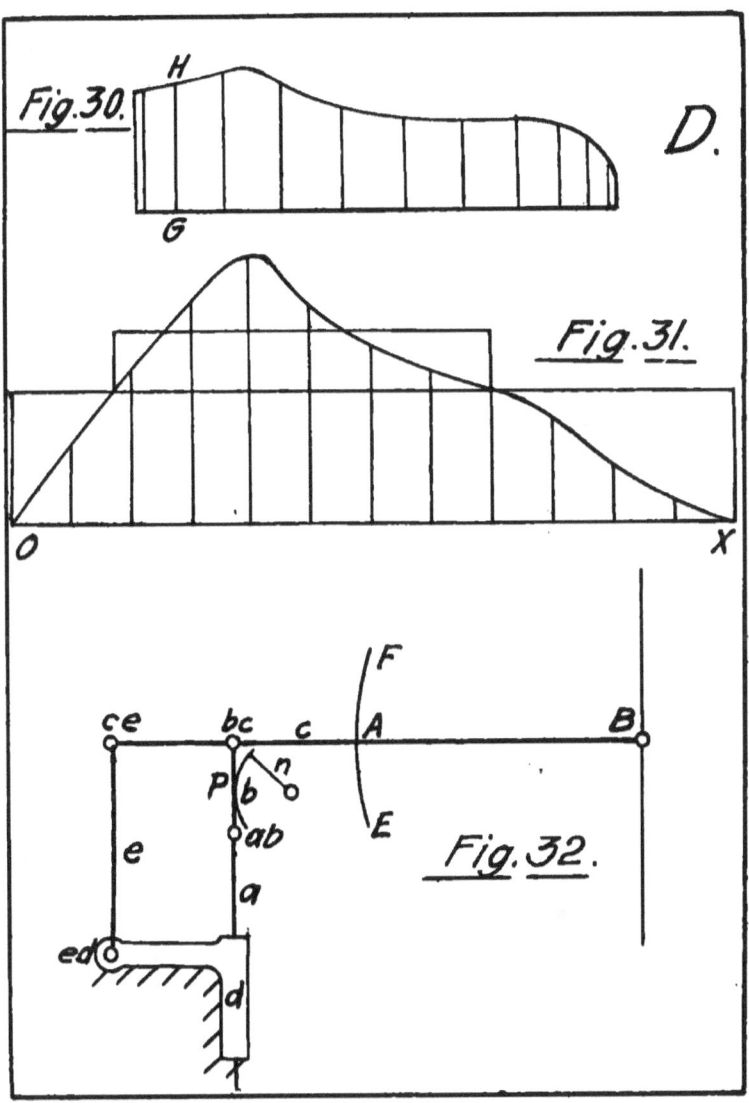

CHAPTER IV.

PARALLEL OR STRAIGHT LINE MOTIONS.

37. Rectilinear motion in machines is usually obtained by means of prismatic guides. It is sometimes necessary, however, to accomplish the same result by linkages.

A general method of design, which is applicable in many cases, will be given. In Fig. 32, d is the fixed link, and a is connected with it by a sliding pair. a, b, c, and e are connected by turning pairs, as shown. The constrainment is not complete because B is free to move in any direction, and its motion would, therefore, depend upon the force producing it. It is required that the point B shall move in a straight line parallel to a. Suppose that B is caused to move along the required line; then any point of the link c, as A, will describe some curve, FAE. If a pin be attached to c, with its axis at A, and a curved slot fitting the pin, with its sides parallel to FAE, be attached to d, as in Fig. 33, it follows that B can only move in the required straight line. This is the mechanism of the Tabor Steam Engine Indicator.

The curve described by A might approximate a circular arc whose centre could be located, say at O, Fig. 33. Then the curved slot might be replaced by a link, attached to d and c by turning pairs at O and A. This gives B approximately the required motion. This is the mechanism of the Thompson Steam Engine Indicator.

If, while the point B is caused to move in the required straight line, a point in b, as P, Fig. 32, were chosen, it would be found to describe a curve which would approximate a circular arc, whose centre, O, and radius, $= r$, could be found. Let the link whose length $= r$ be attached to d and b by turning pairs whose axes are at O and P, and the motion of B will be approximately the required motion. This is the mechanism of the Crosby Steam Engine Indicator.

38. *Problem.* — In Fig. 34, B is the fixed axis of a counter-shaft; C is the axis of another shaft which is free to rotate about B. D is the axis of a circular saw which is free to move in any direction. It is required to constrain D to move in the straight line EF. If D be moved along EF, a tracing point fixed at A in the link CD will describe an approximate circular arc, HAK, whose centre may be found at O. A link whose length is OA may be connected to the fixed link, and to the link CD by means of turning pairs at O and A. D will then be constrained to move approximately along EF. A curved slot and pin could be used, and the motion would be exact.*

* Descriptions of many varieties of parallel motions may be found in Rankine's " Machinery and Millwork "; Weisbach's " Mechanics of Engineering," Vol. III, "Mechanics of the Machinery of Transmission "; Kennedy's " Mechanics of Machinery."

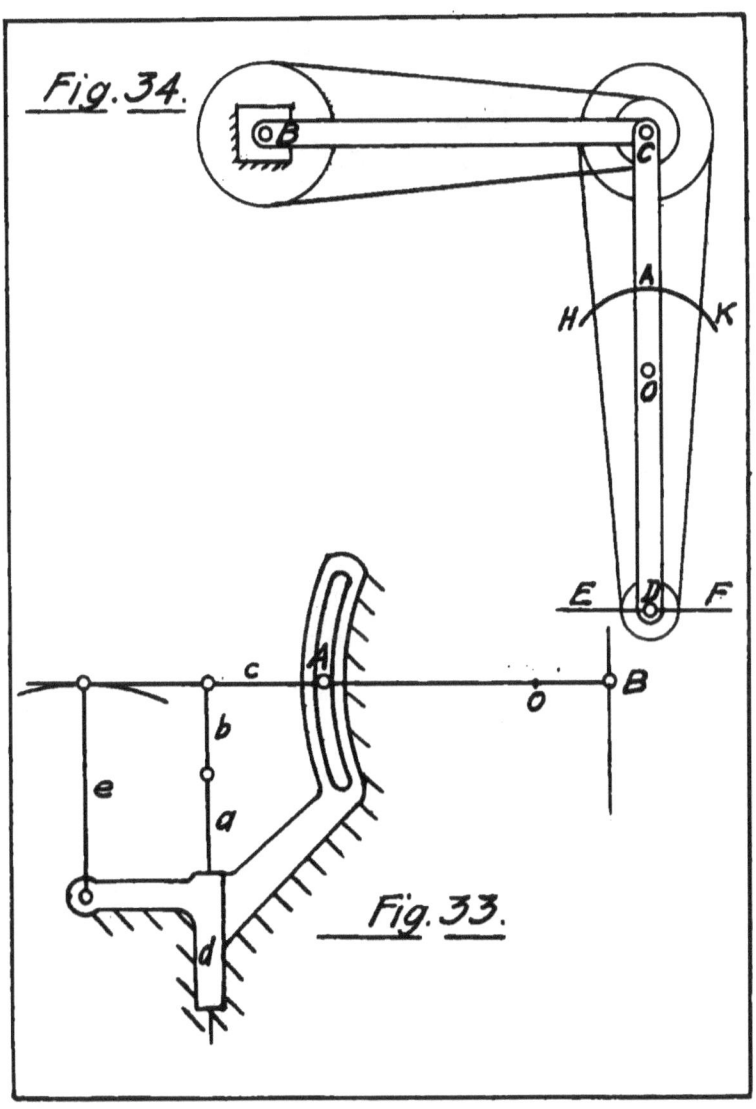

CHAPTER V.

TOOTHED WHEELS, OR GEARS.

39. When toothed wheels are used to communicate motion, the motion elements are the tooth surfaces. The contact of these surfaces with each other is line contact. Such pairs of motion elements are called *higher pairs*, to distinguish them from lower pairs, which are in contact throughout their entire surface. Fig. 35 shows the simplest toothed wheel mechanism. There are three links, a, b, and c, and therefore three centros ab, bc, and ac. These centros must, as heretofore explained, lie in the same straight line. ac and ab are the centres of the turning pairs connecting c and b to a. It is required to locate bc on the line of centres.

When the gear c is caused to rotate uniformly with a certain angular velocity, *i. e.*, at the rate of m revolutions per minute, it is required to cause the gear b to rotate uniformly at a rate of n revolutions per minute. The angular velocity-ratio is therefore constant, and $=\dfrac{m}{n}$. The centro bc is a point on the line of centres which has the same linear velocity whether it is considered as a point in b or c. The linear velocity of this point bc in $b = 2\pi R_1 n$; and the linear velocity of the same point in $c = 2\pi R_2 m$; in which $R_1 =$ radius of bc in b, and $R_2 =$ radius of bc in c. But this linear velocity must be the same in both cases, and hence the above expressions may be equated thus:

$$2\pi R_1 n = 2\pi R_2 m \; ;$$

whence
$$\frac{R_1}{R_2} = \frac{m}{n}.$$

40   MACHINE DESIGN.

Hence bc is located by dividing the line of centres into parts which are to each other inversely as the angular velocities of the gears.

Thus, let ab and ac, Fig. 36, be the centres of a pair of gears whose angular velocity ratio $= \dfrac{m}{n}$. Draw the line of centres; divide into $m + n$ equal parts; m of these from ab toward the right, or n from ac toward the left, will locate bc. Draw circles through bc, with ab and ac as centres. These circles are the centroids of bc and are called *pitch circles*. It has been already explained that any motion may be reproduced by rolling the centroids of that motion upon each other without slipping. Therefore the motion of gears is the same as that which would result from the rolling together of the pitch circles (or cylinders) without slipping. In fact, these pitch cylinders themselves might be, and sometimes are, used for transmitting motion of rotation. Slipping, however, is apt to occur, and hence these "friction gears" cannot be used if no variation from the given velocity ratio is allowable. Hence, teeth are formed on the wheels which engage with each other, to prevent slipping.

40. Teeth of almost any form may be used, and the *average* velocity will be right. But if the forms are not correct there will be continual variations of velocity ratio between a minimum and maximum value. These variations are in many cases unallowable. and in all cases undesirable. It is necessary therefore to study tooth outlines which shall serve for the transmission of a constant velocity ratio.

The centro of relative motion of the two gears must remain in a constant position in order that the velocity ratio shall be constant. *The essential condition for constant velocity ratio is, therefore, that the position of the centro of relative motion of the gears shall remain unchanged.* If A and B, Fig. 38, are tooth surfaces in contact at a, their only possible relative motion, if they remain in contact, is slipping motion along the tangent CD. The centro of this motion must be in EF, a normal to the tooth surfaces at the point of contact. If these be supposed to be teeth of a pair of gears, b and c, whose required velocity ratio is known, and whose centro, bc, is

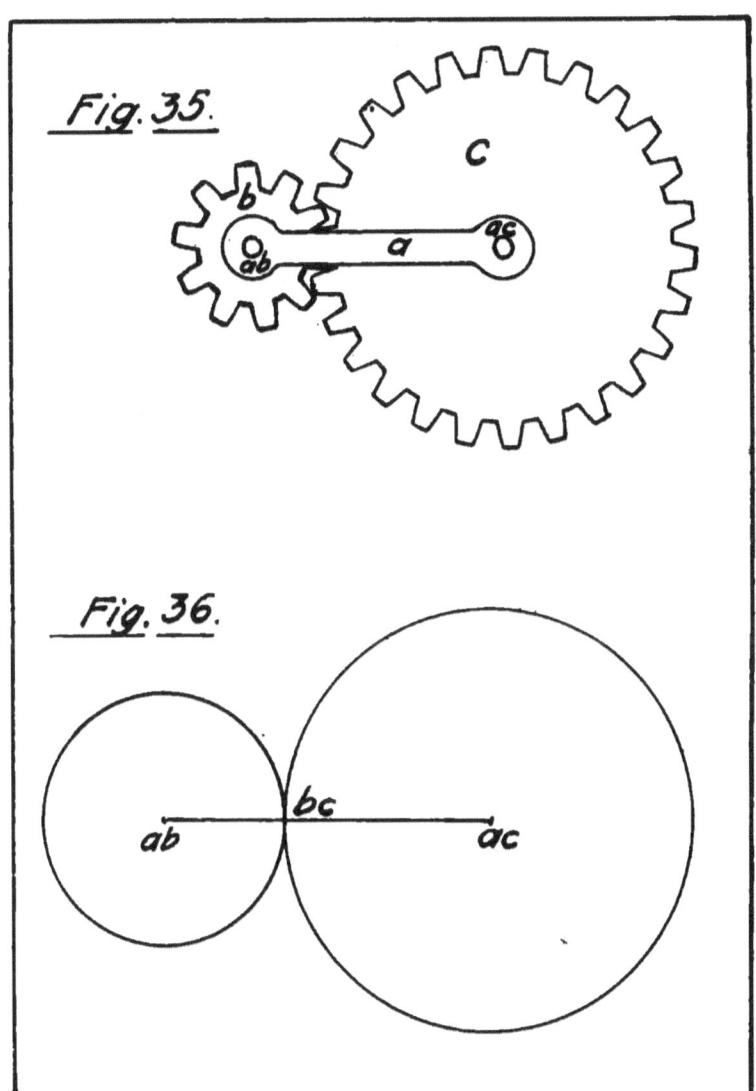

therefore located, then in order that the motion communicated from one gear to the other through the point of contact, a, shall be the required motion, it is necessary that the centro of the relative motion of the teeth shall coincide with bc.

Illustration. — In Fig. 38, let ac and ab be centres of rotation of bodies b and c, and the required velocity ratio is such that the centro of b and c falls at bc. Contact between b and c is at p. The only possible relative motion if these surfaces remain in contact is slipping along CD; hence the centro of this motion must be on EF, the normal to the tooth surfaces at the point of contact. But it must also be on the same straight line with ac and ab; hence it is at bc, and the motion transmitted for the instant, at the point p, is the required motion, because its centro is at bc. But the curves touching at p, might be of such form that their common normal at p would intersect the line of centres at some other point, as K, which would then become the centro of the motion of b and c for the instant, and would correspond to the transmission of a different motion. The essential condition to be fulfilled by tooth outlines, in order that a constant velocity ratio may be maintained, may therefore be stated as follows: *The tooth outlines must be such that their normal at the point of contact shall always pass through the centro corresponding to the required velocity ratio.*

41. Having given any curve that will serve for a tooth outline in one gear, the corresponding curve may be found in the other gear, which will engage with the given curve and transmit a constant velocity ratio. Let $\frac{m}{n}$ be the given velocity ratio. Draw the line of centres AB, Fig. 39. Let P be the "pitch point," *i. e.*, the point of contact of the pitch circles or the centro of relative motion of the two gears. To the right from P lay off a distance $PB = m$; from P toward the left lay off $PA = n$. A and B will then be the required centres of the wheels, and the pitch circles may be drawn through P. Let abc be *any* given curve on the wheel A. It is required to find the curve in B which shall engage with abc to transmit the constant velocity ratio required. A normal to the point of contact must pass through the centro. If, therefore, any point, as a, be taken in the given curve, and a normal to the curve at that point be drawn, as aa, then when a is the point of contact, a will coincide with P. Also, if cr is a normal to the curve at c, then r will coincide with P when c is the point of contact between the gears; and since b is in the pitch line, it will itself coincide with P when it is the point of contact. Suppose now that A and B are discs of cardboard, that A overlaps B, and that a thread is stretched to indicate the centre line AB. Suppose also that they can be rotated so that the pitch circles roll on each other without slipping. Roll the discs till a reaches P, and prick a through upon B; then make b coincide with P, and prick it through; then make r coincide with P, and prick c through. This will give three points in the required curve in B, and through these the curve may be drawn. The curve could, of course, be more accurately located by using more points. The points of the curve in B might be located geometrically.

Many curves could be drawn that would not serve for tooth outlines; but, given any curve that will serve, the corresponding curve may be found. There would be, therefore, almost an infinite number of curves, that would fulfill the requirements of correct tooth outlines. But in practice two kinds of curves are found so convenient that they are most commonly, though not exclusively, used. They are *cycloidal* and *involute* curves.

42. It is assumed that the character of cycloidal curves and method of drawing them is understood.

In Fig. 40, let b and c be the pitch circles of a pair of wheels, always in contact at bc. Also let m be the describing circle in contact with both at the same point. M is the describing point. When one curve rolls upon another, the centro of their relative motion is always their point of contact. For, since the motion of rolling excludes slipping, the two bodies must be stationary, relatively to each other, at their point of contact; and bodies that move relatively to each other can have but one such stationary point in common — their centro. When, therefore, m rolls in or

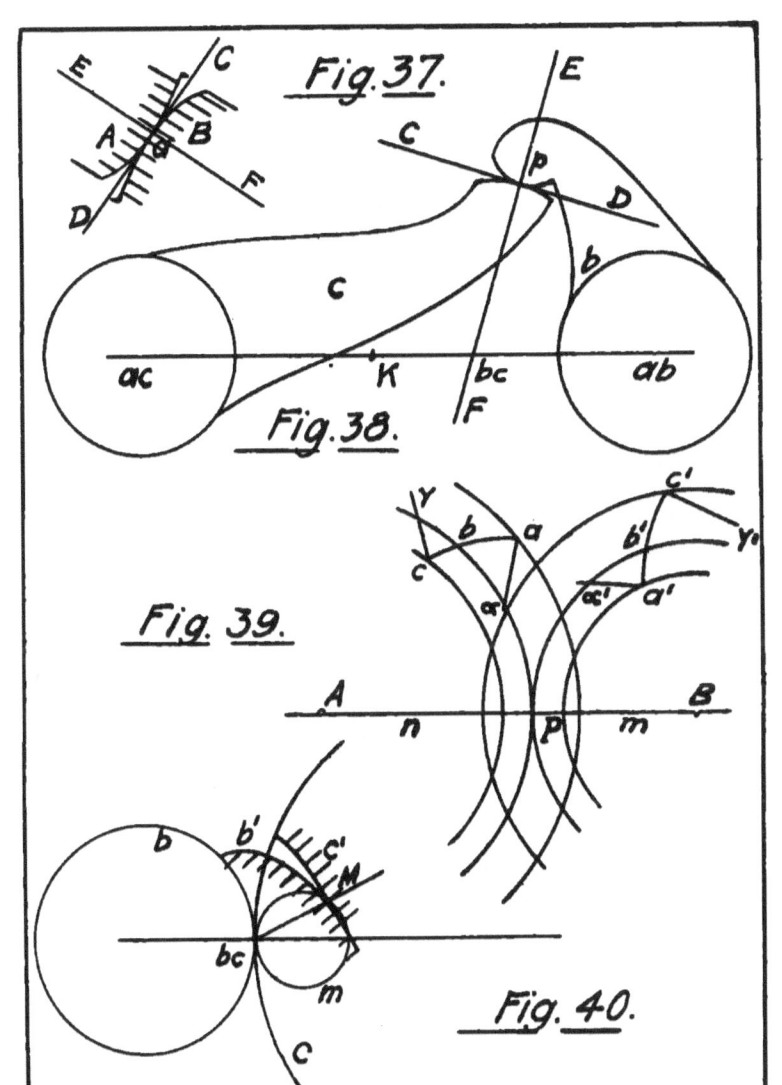

upon b or c, its centro relatively to either is their point of contact. The point M, therefore, must describe curves whose direction at any point is at right angles to a line joining that point to the point of contact of m with the circle. Suppose the two circles b and c to revolve about their centres, being always in contact at bc; suppose m to rotate at the same time, the three circles being always in contact at one point. The point M will then describe simultaneously a curve, b', on the plane of b, and a curve, c', on the plane of c. Since M describes the curves simultaneously, it will always be the point of contact between them in any position. And since the point M moves always at right angles to a line which joins it to bc, therefore the normal to the tooth surfaces at their point of contact will always pass through bc, and the condition for constant velocity ratio transmission is fulfilled. But these curves are precisely the epicycloid and hypocycloid that would be drawn by the point M in the generating circle, by rolling on the outside of b and inside of c. Obviously, then, the epicycloids and hypocycloids generated in this way, used as tooth profiles, will transmit a constant velocity ratio.

This proof is independent of the size of the generating circle, and its diameter may therefore equal the radius of b. Then the hypocycloids generated by rolling within b would be straight lines coinciding with the radius of b. In this case the profiles of the teeth of b become radial lines; and therefore the teeth are thinner at the base than at the pitch line; for this reason they are weaker than if a smaller generating circle had been used. All tooth curves generated with the same generating circle will work together, the pitch being the same. It is therefore necessary to use the same generating circle for a set of gears which need to *interchange*.

The describing circle may be made still larger. In the first case the curves described have their convexity in the same direction; *i. e.*, they lie on the same side of a common tangent. When the diameter of the describing circle is made equal to the radius of b, one curve becomes a straight line tangent to the other curve. As the describing circle becomes still larger, the curves have their convexity in opposite directions. As the circle approximates equality with b, the hypocycloid grows shorter, and finally, when the describing circle equals b, it becomes a point which is the generating point in b, which is now the generating circle. If this point could be replaced by a pin having no sensible diameter, it would engage with the epicycloid generated by it in the other gear to transmit a constant velocity ratio. But a pin without sensible diameter will not serve as a wheel tooth, and a proper diameter must be assumed, and a new curve laid off to engage with it in the other gear. In Fig. 41, AB is the epicycloid generated by a point in the circumference of the other pitch circle. CD is the new curve drawn tangent to a series of positions of the pin as shown. The pin will engage with this curve, CE, and transmit the constant velocity ratio as required. In Fig. 40, let it be supposed that when the three circles rotate constantly tangent to each other at the pitch point bc, a pencil is fastened at the point M in the circumference of the describing circle. If this pencil be supposed to mark simultaneously upon the planes of b, c, and that of the paper, it will describe upon b an epicycloid, on c a hypocycloid, and on the plane of the paper an arc of the describing circle. Since M is always the point of contact of the cycloidal curves (because it generates them simultaneously), therefore, in cycloidal gear teeth, *the locus or path of the point of contact is an arc of the describing circle.*

43. In the cases already considered, where an epicycloid in one wheel engages with a hypocycloid in the other, the contact of the teeth with each other is all on one side of the line of centres. Thus, in Fig. 40, if the motion be reversed, the curves will be in contact until M returns to bc along the arc MD-bc; but after M passes bc contact will cease. If c were the driving wheel, the point of contact would approach the line of centres; if b were the driving wheel the point of contact would recede from the line of centres. Experience shows that the latter gives smoother running because of better conditions as regards friction between the tooth surfaces. It would be desirable, therefore, that the wheel with the epicycloidal curves should always be the driver. But it should be possible to use either wheel as driver to meet the varying conditions in practice.

Another reason why contact should not be all on one side of the line of centres may be explained as follows:

Definitions. — The angle through which a gear wheel turns, while one of its teeth is in contact with the corresponding tooth in the other gear, is called *the angle of action*. The arc of the pitch circle corresponding to the angle of action is called *the arc of action*.

The arc of action must be greater than the "pitch arc" (the arc of the pitch circle that includes one tooth and one space), or else contact will cease between one pair of teeth before it begins between the next pair. Constrainment would therefore not be complete.

In Fig. 42, let AB and CD be the pitch circles of a pair of gears, and E the describing circle. Let an arc of action be laid off on each of the circles from P, as Pa, Pc, and Pe. Through e, about the centre O, draw an addendum circle, *i. e.*, the circle which limits the points of the teeth. Since the circle E is the path of the point of contact, and since the addendum circle limits the points of the teeth, their intersection, e, is the point at which contact ceases, rotation being as indicated by the arrow. If the pitch arc just equals the assumed arc of action, contact will be just beginning at P when it ceases at e; but if the pitch arc be greater than the arc of action, contact will not begin at P till after it has ceased at e, and there will be an interval when AB will not drive CD. The greater the arc of action the greater the distance of e from P on the circumference of the describing circle. The direction of pressure between the teeth is always a normal to the tooth surface, and this always passes through the pitch point; therefore, the greater the arc of action, *i.e.*, the greater the distance of e from P, the greater the obliquity of the line of pressure. The pressure may be resolved into two components, one at right angles to the line of centres, and the other parallel to it. The first is resisted by the teeth of the follower wheel, and therefore produces rotation; the second is resisted at the journal, and produces pressure, with resulting friction. Hence, it follows that the greater the arc of action, the greater will be the average obliquity of the line of pressure, and therefore the greater the component of the pressure that produces wasteful friction. If it can be arranged so that the arc of action shall be partly on each side of the line of centres, the arc of action may be made greater than the pitch arc (usually equal to about 1½ times the pitch arc); then the obliquity of the pressure line may be kept within reasonable limits, contact between the teeth will be insured in all positions, and either wheel may be the driver. This is accomplished by using two describing circles as in Fig. 43. Suppose the four circles A, B, a, and β, to roll constantly tangent at P. a will describe an epicycloid on the plane of B, and a hypocycloid on the plane of A. These curves will engage with each other to drive correctly. β will describe an epicycloid on A, and a hypocycloid on B. These curves will engage also, to drive correctly. If the epi- and hypocycloid in each gear be drawn through the same point on the pitch circle, a double curve tooth outline will be located, and one curve will engage on one side of the line of centres, and the other on the other side. If A drives as indicated by the arrow, contact will begin at D, and the point of contact will follow an arc of a to P, and then an arc of β to C.

44. Involute Tooth Outlines. — If a string be wound around a cylinder and a pencil point attached to its end, this point will trace an involute, as the string is unwound from the cylinder. Or, if the point be constrained to follow a tangent to the cylinder, and the string be unwound by rotating the cylinder about its axis, the point will trace an involute on a plane that rotates with the cylinder. *Illustration.* — Let a, Fig. 44, be a circular piece of wood, free to rotate about C; β is a circular piece of cardboard made fast to a; AB is a straight-edge held on the circumference of a, having a pencil point at B. As B moves along the straight-edge to A, a and β rotate about C, and B traces an involute DB upon β. The relative motion of the tracing point and β being just the same as if the string had been simply unwound from a, fixed. If the tracing point is caused to return along the straight-edge it will trace the involute BD in a reverse direction.

The centro of the tracing point is always the point of tangency of the string with the cylinder; therefore the string, or straight-

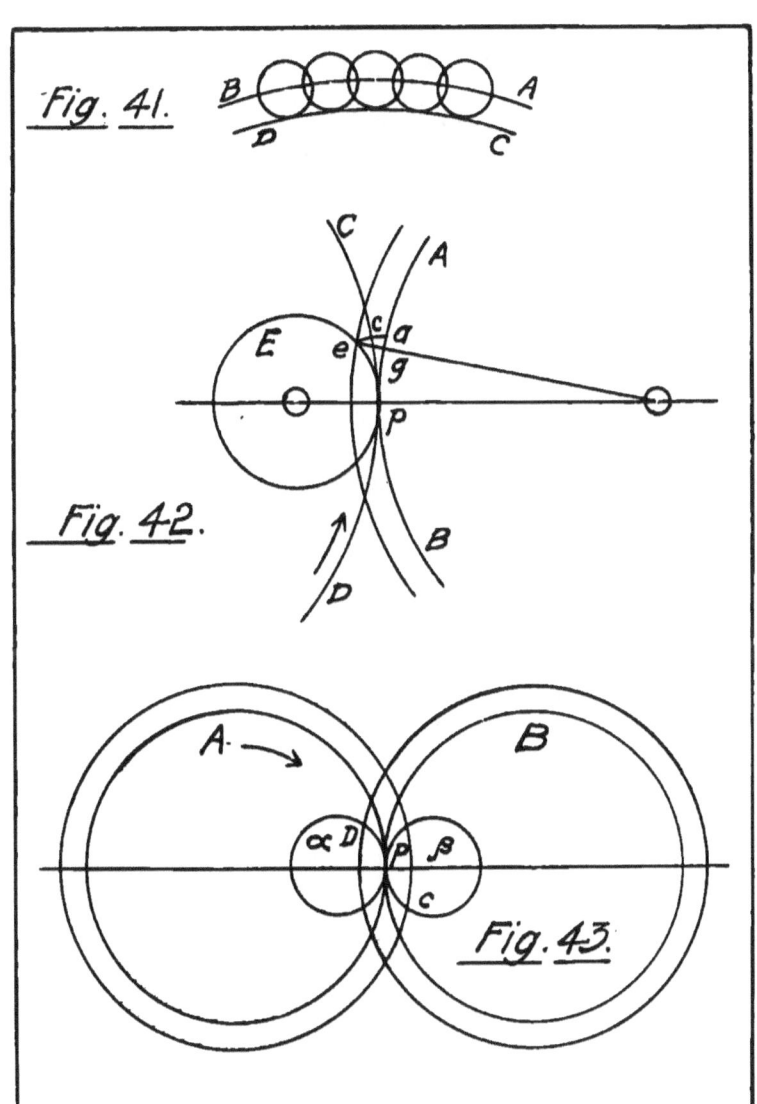

edge, in Fig. 45, is always at right angles to the direction of motion of the tracing point, and hence is always a normal to the involute curve. Let a and β, Fig. 45, be two base cylinders; let AB be a cord wound upon a and β and passing through the centro P, which corresponds to the required velocity ratio. Let a and β be supposed to rotate so that the cord is wound from β upon a. Then any point in the cord will move from A toward B, and, if it be a tracing point, will trace an involute of β on the plane of β (extended beyond the base cylinder), and will also trace an involute of a upon the plane of a. These two involutes will serve for tooth profiles for the transmission of the required constant velocity ratio, because AB is the constant normal to both curves at their point of contact, and it passes through P, the centro corresponding to the required velocity ratio. Hence, the necessary condition is fulfilled.

Since a point in the line AB describes involute curves simultaneously, the point of contact of the curves is always in the line AB. And hence AB is the path of the point of contact.

One of the advantages of involute curves for tooth profiles is that a change in distance between centres of the gears, does not interfere with the transmission of a constant velocity ratio. This may be proved as follows: In Fig. 45, from similar triangles $\frac{OB}{O'A} = \frac{OP}{O'P}$; that is, the ratio of the radii of the base circles is equal to the ratio of the radii of the pitch circles. This ratio equals the inverse ratio of angular velocities of the gears. Suppose now that O and O' be moved nearer together: the pitch circles will be smaller, but the ratio of their radii must be equal to the unchanged ratio of the radii of the base circles, and therefore the velocity ratio remains unchanged. Also the involute curves, since they are generated from the same base cylinders, will be the same as before, and therefore, with the same tooth outlines, the same constant velocity ratio will be transmitted as before.

45. Definitions. — If the pitch circle be divided into as many equal parts as there are teeth in the gear, the arc included between two of these divisions is the **circular pitch** of the gear. Circular pitch may also be defined as the distance on the pitch circle occupied by a tooth and a space; or, otherwise, it is the distance on the pitch circle from any point of a tooth to the corresponding point in the next tooth. A fractional tooth is impossible, and therefore the circular pitch must be such a value that the pitch circumference is divisible by it. Let P = circular pitch in inches; let D = pitch diameter in inches; N = number of teeth; then $NP = \pi D$; $N = \dfrac{\pi D}{P}$; $D = \dfrac{NP}{\pi}$; $P = \dfrac{\pi D}{N}$. From these relations any one of the three values, P, D, and N, may be found if the other two are given.

Diametral pitch is the number of teeth per inch of pitch diameter. Thus, if p = diameter pitch, $p = \dfrac{N}{D}$. Multiplying the two expressions, $P = \dfrac{\pi D}{N}$ and $p = \dfrac{N}{D}$, together, gives $Pp = \dfrac{\pi D}{N} \cdot \dfrac{N}{D} = \pi$. Or, the product of diametral and circular pitch $= \pi$. Circular pitch is usually used for large cast gears, and for mortice gears (gears with wooden teeth inserted). Diametral pitch is usually used for small cut gears.

In Fig. 46, b, e, and k, are pitch points of the teeth; ab is the **face** of the tooth; bm is the **flank** of the tooth; AD is the **total depth** of the tooth; AC is the **working depth**; AB is the **addendum**; a circle through A is the **addendum** circle. **Clearance** is the excess of total depth over working depth, $= CD$. **Backlash** is the width of space on the pitch line, minus the width of the tooth on the same line. In cast gears whose tooth surfaces are not "tooled" backlash needs to be allowed, because of unavoidable imperfections in the surfaces. In cut gears, however, it may be reduced almost to zero, and the tooth and space, measured on the pitch circle, may be considered equal.

46. Racks. — A rack is a wheel whose pitch radius is infinite; its pitch circle, therefore, becomes a straight line, and is tangent to the pitch circle of the wheel, or pinion with which the rack engages. The line of centres is a normal to the pitch line of the

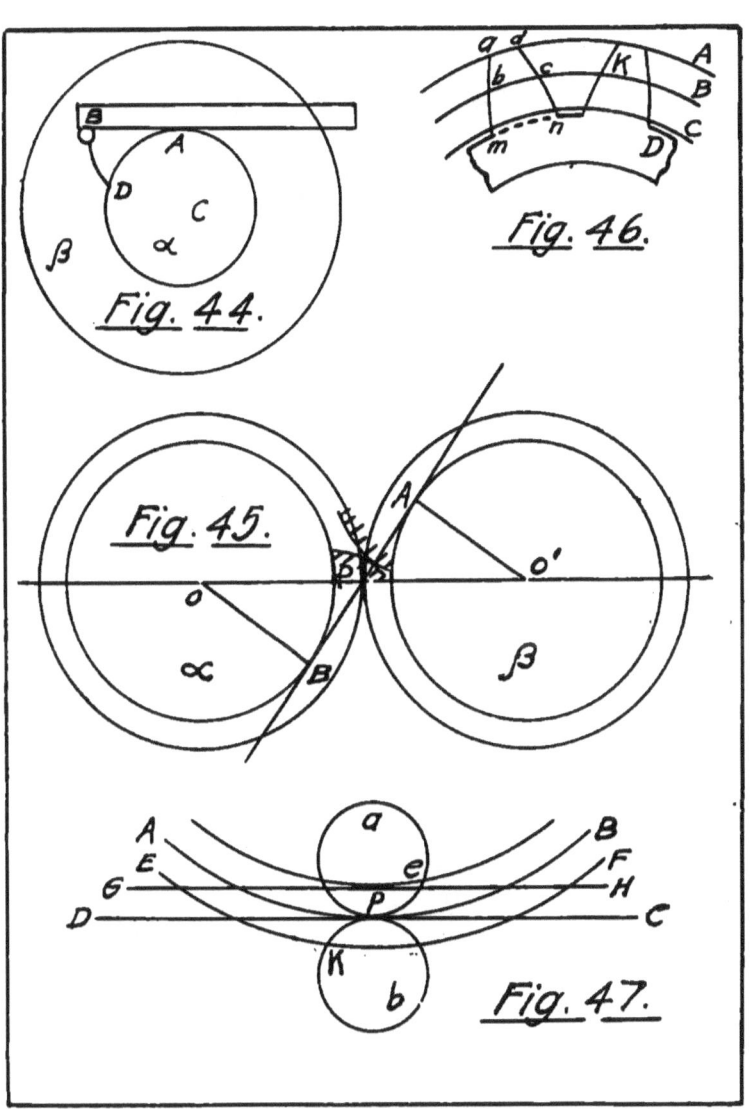

rack, through the centre of the pitch circle of the pinion. The pitch of the rack is determined by laying off the circular pitch of the engaging wheel on the pitch line of the rack. The curves of the rack teeth, like those of wheels of finite radius, may be generated by a point in the circumference of a circle which rolls on the pitch circle. Since, however, the pitch circle is now a straight line, the tooth curves will be cycloids, both for flanks and faces. In Fig. 47, AB is the pitch circle of the pinion, and CD is the pitch line of the rack; a and b are describing circles. Suppose, as before, that all move without slipping, and are constantly tangent at P. A point in the circumference of a will then describe simultaneously a cycloid on CD, and a hypocycloid within AB. These will be correct tooth outlines. Also, a point in the circumference of b will describe a cycloid on CD and an epicycloid on AB. These will be correct tooth outlines. To find the path of the point of contact, draw the addendum circle EF of the pinion, and the addendum line GH of the rack. When the pinion turns clockwise and drives the rack, contact will begin at e and follow arcs of the describing circles through P to K. It is obvious that a rack cannot be used where rotation is continuous in one direction, but only where motion is reversed.

Involute curves may also be used for the outlines of rack teeth. Let CD and $C'D'$, Fig. 48, be the pitch lines. When it is required to generate involute curves for tooth outlines, for a pair of gears of finite radius, a line is drawn through the pitch point at a given angle to the line of centres (usually 75°); this line is the path of the point which generates two involutes simultaneously, and therefore the path of the point of contact between the tooth curves. It is also the common tangent to the two base circles, which may now be drawn and used for the describing of the involutes. To apply this to the case of a rack and pinion, draw EF, Fig. 48. The base circles must be drawn tangent to this line; AB will therefore be the base circle for the pinion. But the base circle in the rack has an infinite radius, and a circle of infinite radius drawn tangent to EF would be a straight line coincident with EF. Therefore EF is the base line of the rack. But an involute to a base circle of infinite radius is a straight line normal to the circumference — in this case a straight line perpendicular to EF. Therefore the tooth profiles of a rack in the involute system will always be straight lines perpendicular to the path of the describing point, and passing through the pitch points. If, in Fig. 48, the pinion move clockwise and drive the rack, the contact will begin at E, the intersection of the addendum line of the rack GH and the base circle AB of the pinion, and will follow the line EF through P to the point where EF cuts the addendum circle LM of the pinion.

47. Annular Gears. — Both centres of a pair of gears may be on the same side of the pitch point. This arrangement corresponds to what is known as an annular gear and pinion. Thus, in Fig. 49, AB and CD are the pitch circles, and their centres are both above the pitch point P. Teeth may be constructed to transmit rotation between AB and CD. AB will be an ordinary spur pinion, but it is obvious that CD becomes a ring of metal with teeth on the inside, $i.e.$, it is an annular gear. In this case a and β may be describing circles, and a point in the circumference of a will describe hypocycloids simultaneously on the planes of AB and CD; and a point in the circumference of β will describe epicycloids simultaneously on the planes of AB and CD. These will engage to transmit a constant velocity ratio. Obviously the space inside of an annular gear corresponds to a spur gear of the same pitch and pitch diameter, with tooth curves drawn with the same describing circle. Let EF and GH, Fig. 49, be the addendum circles. If the pinion move clockwise, driving the annular gear, the path of the point of contact will be from e along the circumference of a to P, and from P along the circumference of β to K.

The construction of involute teeth for an annular gear and pinion involves exactly the same principle as in the case of a pair of spur gears. The only difference of detail is that the describing point is in the tangent to the base circles *produced* instead of being between the points of tangency. Let O and O', Fig. 50, be the centres, and AB and IJ the pitch circles of an annular gear

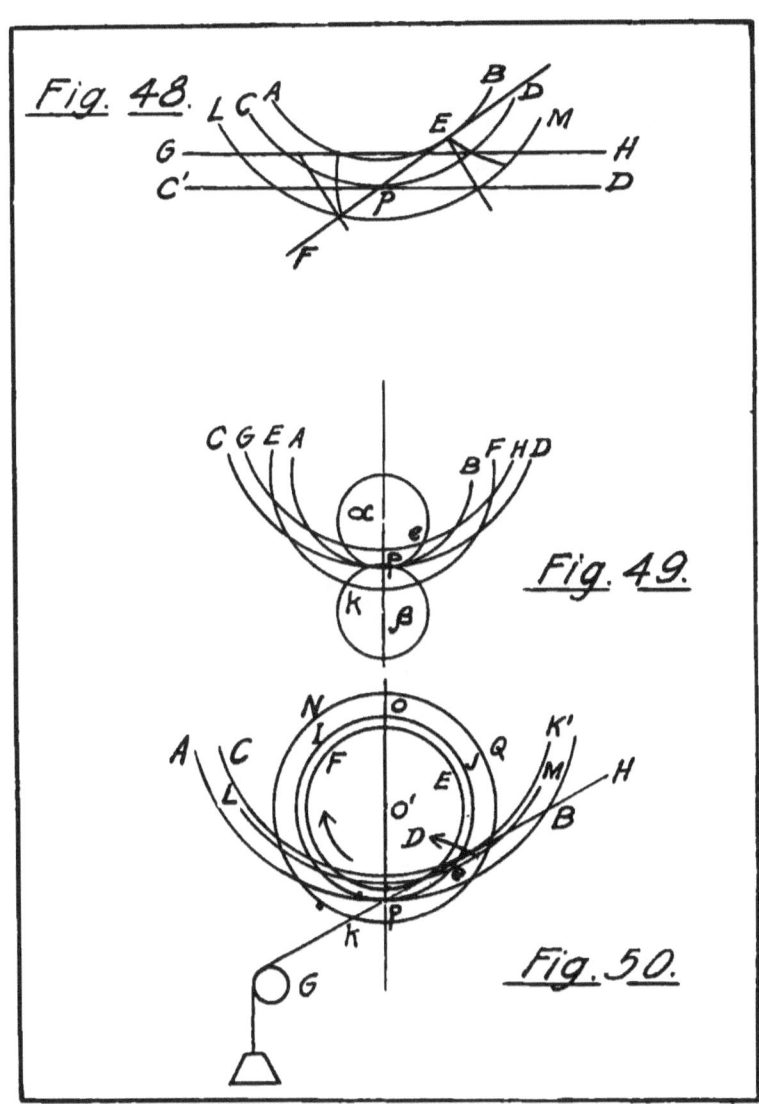

and pinion. Through P, the point of tangency of the pitch circles, draw the path of the point of contact, at the given angle with the line of centres. With O and O' as centres draw tangent circles to this line. These will be the involute base circles. Let the tangent be replaced by a cord, made fast say at K', winding on the circumference of the base circle CK', to D, and then around the base circle FE in the direction of the arrow, and passing over the pulley G which holds it in line with PB. If rotation be supposed to occur with the two pitch circles always tangent at P without slipping, any point in the cord beyond P toward G, will describe an involute on the plane IJ, and another on the plane of AB. These will be the correct involute tooth profiles required. Draw NQ and LM, the addendum circles. Then if the pinion move clockwise, driving the annular gear, the point of contact starts from e and moves along the line GH through P to K.

When a pair of spur gears mesh with each other, the direction of rotation is reversed. But an annular gear and pinion meshing together, rotate in the same direction.

48. Interchangeable Sets of Gears. — In practice it is desirable to have "interchangeable sets" of gears; *i. e.*, sets in which any gear will "mesh" correctly with any other, from the smallest pinion to the rack, and in which, except for limiting conditions of size, any spur gear will mesh with any annular gear. Interchangeable sets may be made in either the cycloidal or involute system. A necessary condition in any set is, that the *pitch shall be constant;* because the thickness of tooth on the pitch line must always equal the width of the space (less clearance). If this condition is unfulfilled they cannot engage, whatever the form of the tooth outlines.

The second condition for an interchangeable set in the cycloidal system is that the *size of the describing circle shall be constant.* If the diameter of the describing circle equal the radius of the smallest pinion's pitch circle, the flanks of this pinion's teeth will be radial lines, and the tooth will therefore be thinner at the base than at the pitch line. As the gears increase in size with this constant size of describing circle, the teeth grow thicker at the base; hence, the weakest teeth are those of the smallest pinion.

It is found unadvisable to make a pinion with less than twelve teeth. If the radius of a fifteen-tooth pinion be selected for the diameter of the describing circle, the flanks in a twelve-tooth pinion will be very nearly parallel, and may therefore be cut with a milling cutter. This would not be possible if the describing circle were made larger, causing the space to become wider at the bottom than at the pitch circle. Therefore the maximum describing circle for milled gears is one whose diameter equals the pitch radius of a fifteen-tooth pinion, and it is the one usually selected. Each change in the number of teeth with constant pitch causes a change in the size of the pitch circle. Hence, the form of the tooth outline, generated by a describing circle of constant diameter, also changes. For any pitch, therefore, a separate cutter would be required corresponding to every number of teeth, to insure absolute accuracy. Practically, however, this is not necessary. The change in the form of tooth outline is much greater in a small gear, for any increase in the number of teeth, than in a large one. It is found that twenty-four cutters will cut all possible gears of any pitch with sufficient practical accuracy. The range of these cutters is indicated in the following table, taken from Brown & Sharpe's "Treatise on Gearing":

Cutter A cuts 12 teeth.   Cutter M cuts 27 to 29 teeth.
"   B "   13   "   "   N "   30 to 33   "
"   C "   14   "   "   O "   34 to 37   "
"   D "   15   "   "   P "   38 to 42   "
"   E "   16   "   "   Q "   43 to 49   "
"   F "   17   "   "   R "   50 to 59   "
"   G "   18   "   "   S "   60 to 74   "
"   H "   19   "   "   T "   75 to 99   "
"   I "   20   "   "   U "   100 to 149 "
"   J "   21 to 22 teeth. " V "   150 to 249 "
"   K "   23 to 24   "   " W "   250 to rack.
"   L "   24 to 26   "   " X "   rack.

These same principles of interchangeable sets of gears, with cycloidal tooth outlines, apply not only to small milled gears as above, but also to large cast gears with tooled or untooled tooth surfaces.

49. Interchangeable Involute Gears. — In the involute system the second condition of interchangeability is that the *angle between the common tangent to the base circles and the line of centres shall be constant.* This may be shown as follows: Draw the line of centres, AB, Fig. 51. Through P, the assumed pitch point, draw CD, and let it be the constant common tangent to all base circles from which involute tooth curves are to be drawn. Draw any pair of pitch circles tangent at P, with their centres in the line AB. About these centres draw circles tangent to CD; these are base circles, and CD may represent a cord that winds from one upon the other. A point in this cord will generate, simultaneously, involutes that will engage for the transmission of a constant velocity ratio. But this is true of *any* pair of circles that have their centres in AB, and are tangent to CD. Therefore, if the pitch is constant, any pair of gears that have the base circles tangent to the line CD, will mesh together properly. As in the cycloidal gears, the involute tooth curves vary with a variation in the number of teeth, and, for absolute theoretical accuracy, there would be required for each pitch as many cutters as there are gears with different numbers of teeth. The variation is least at the pitch line, and increases with the distance from it. The involute teeth are usually used for the finer pitches, and the cycloidal teeth for the coarser pitches; and since the amount that the tooth surface extends beyond the pitch line increases with the pitch, it follows that the variation in form of tooth curves is greater in the coarse pitch cycloidal gears than in the fine pitch involute gears. For this reason, with involute gears, it is only necessary to use *eight* cutters for each pitch. The range is shown in the following table, which is also taken from Brown & Sharpe's "Treatise on Gearing":

No. 1 will cut wheels from 135 teeth to racks.
" 2 "   "   "   "   55 " to 134 inclusive.
" 3 "   "   "   "   35 " to 54   "
" 4 "   "   "   "   26 " to 34   "
" 5 "   "   "   "   21 " to 25   "
" 6 "   "   "   "   17 " to 20   "
" 7 "   "   "   "   14 " to 16   "
" 8 "   "   "   "   12 " to 13   "

50. Laying Out Gear Teeth. Exact and Approximate Methods.— There is ordinarily no reason why an exact method for laying out cycloidal or involute curves for tooth outlines should not be used, either for large gears or gear patterns, or in making drawings. It is required to lay out a cycloidal gear. The pitch, and diameters of pitch, and describing circle are given.—Draw the pitch circle. From a piece of thin wood cut out a template to fit a segment of the pitch circle from the inside, as A, Fig. 52. Cut another template to fit a segment of the pitch circle from the outside, as B. Also cut a wooden disc whose diameter equals that of the given describing circle, and fix a tracing point in its circumference. Divide the pitch circle into parts equal to the given circular pitch. Let P be one of the pitch points. Locate A so that its curved edge coincides with the pitch circle at the right of P. Roll the describing circle on A, without slipping, so that the epicycloid described by the tracing point shall pass through P. Next place B so that its curved edge coincides with the pitch circle at the left of P, and roll the circle on the inside of B, without slipping, so that the hypocycloid described by the tracing point shall pass through P. Thus the outline of one tooth is drawn, aPb. Cut a wooden template to fit the tooth curve, and make it fast to a wooden arm free to rotate about O, making the edge of the template coincide with aPb. It may now be swung successively to the other pitch points, and the tooth outline may be drawn by the template edge. This gives one side of all of the teeth. The arm may now be turned over and the other sides of the teeth may be drawn similarly.

TOOTHED WHEELS, OR GEARS. 55

51. It is required to lay out exact involute teeth. The pitch, pitch circle diameter, and angle of the common tangent are given. — Draw the pitch circle, Fig. 53, and the line of centres AB. Through the pitch point, P, draw CD, the common tangent to the base circles, making the angle β with the line of centres. Draw the base circle about O, tangent to CD. Cut a wooden template to fit the base circle from the inside, as EF; wind on this template a fine cord carrying a pencil at its end, and then unwind this, allowing the pencil to trace an involute curve, ab, which will be a correct tooth form. Let a template, cut to fit this involute, be attached to an arm free to rotate about O, and the tooth outlines may be drawn as before. The bottom of the spaces between the teeth may fall within the base circle, in which case the involute curves are extended inward by radial lines.*

52. The following formulas are given to assist in the practical proportioning of gears :

Let $D =$ pitch diameter.
" $D_1 =$ outside diameter.
" $D_2 =$ diameter of a circle through the bottom of spaces.
" $P =$ circular pitch $=$ space on the pitch circle occupied by a tooth and a space.
" $p =$ diametral pitch $=$ number of teeth per inch of pitch circle diameter.
" $N =$ number of teeth.
" $t =$ thickness of tooth on pitch line.
" $a =$ addendum.
" $c =$ clearance.
" $d =$ working depth of spaces.
" $d_1 =$ full depth of spaces.

* Approximate tooth outlines may be drawn by the use of instruments, such as the Willis odontograph, which locates the centres of approximate circular arcs; the templet odontograph, invented by Prof. S. W. Robinson; or by some geometrical or tabular method for the location of the centres of approximate circular arcs. For descriptions see "Elements of Mechanism," Willis; "Kinematics," McCord; "Teeth of Gears," George B. Grant; "Treatise on Gearing," published by Brown & Sharpe.

56   MACHINE DESIGN.

Then, $D_1 = \dfrac{N+2}{p}$; $D_2 = D - 2(a+c)$;

$N = \dfrac{D\pi}{P}$; $P = \dfrac{D\pi}{N}$; $D = \dfrac{PN}{\pi}$; $N = Dp$;

$p = \dfrac{N}{D}$; $D = \dfrac{N}{p}$; $Pp = \pi$;

$p = \dfrac{\pi}{P}$; $P = \dfrac{\pi}{p}$; $t = \dfrac{P}{2} = \dfrac{\pi}{2p}$, no backlash.

$c = \dfrac{t}{10} = \dfrac{P}{20} = \dfrac{\pi}{p20}$; $d = 2a$; $d_1 = 2a + c$; $a = \dfrac{1}{p}$ inches.

The following dimensions are given as a guide; they may be varied as conditions of design require: Width of face = about $3P$; thickness of rim = $1.25 t$; thickness of arms = $1.25 t$; no taper. The rim may be reinforced by a rib, as shown in Fig. 54. Diameter of hub = $2 \times$ diameter of shaft. Length of hub = width of face + $\frac{1}{2}''$; width of arm at junction with hub = $\frac{1}{6}$ circumference of the hub, for six arms. Make arms taper about $\frac{3}{8}''$ per foot on each side.

53. Strength of Gear Teeth. — The maximum work transmitted by a shaft per unit time may usually be accurately estimated; and, if the rate of rotation is known, the torsional moment may be found. Let O, Fig. 55, represent the axis of a shaft perpendicular to the paper. Let A = maximum work to be transmitted per minute; let N = revolutions per minute; let Fr = torsional moment. Then F is the force factor of the work transmitted, and $2\pi rN$ is the space factor of the work transmitted. Hence, $2F\pi rN = A$, and Fr = torsional moment $= \dfrac{A}{2\pi N}$.

If the work is to be transmitted to another shaft by means of a spur gear whose radius is r_1, then for equilibrium $F_1 r_1 = Fr$, and $F_1 = \dfrac{Fr}{r_1}$. F_1 is the force at the pitch surface of the gear whose radius is r_1, i. e., it is the force to be sustained by the gear teeth. Hence, in general, *the force sustained by the teeth of a gear equals the torsional moment divided by the pitch radius of the gear.*

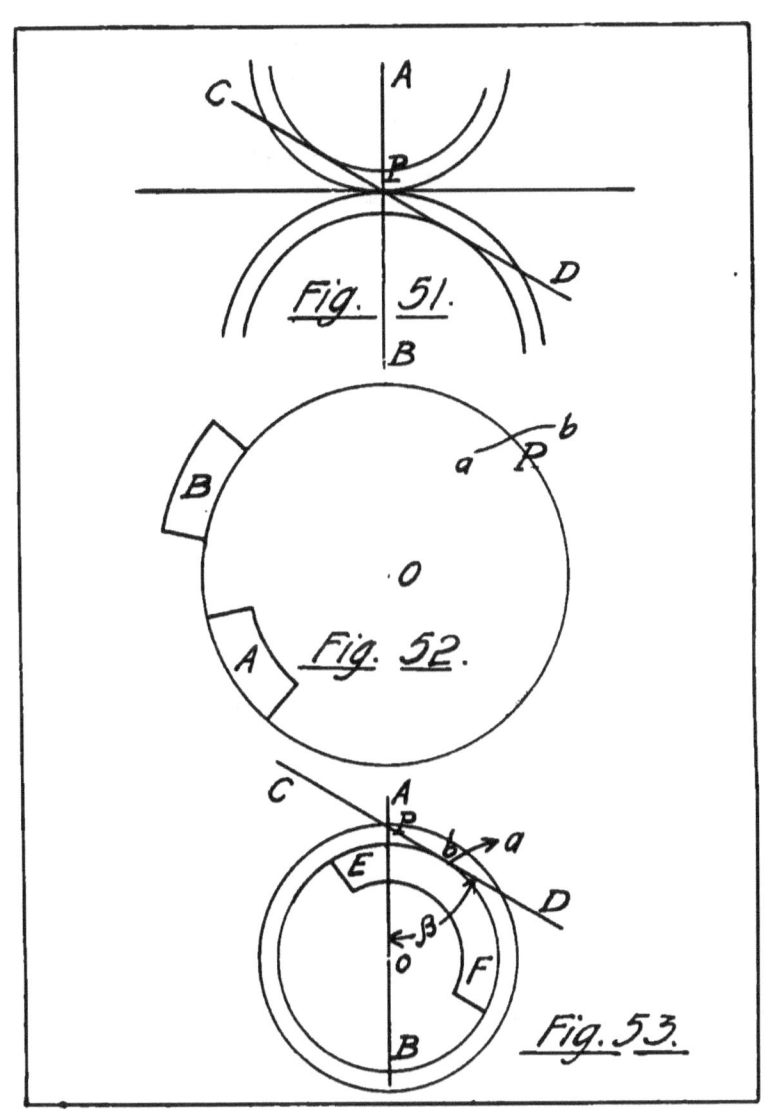

When the maximum force to be sustained is known the teeth may be given proper proportions. The dimensions upon which the tooth depends for strength are : Thickness of tooth $= t$; width of face of gear $= b$; and depth of space between teeth $= l$. These all become known when the pitch is known, because t is fixed for any pitch, and l and b have values dictated by good practice. The value of b may be varied through quite a range to meet the requirements of any special case.

54. In the design the tooth will be treated as a cantilever with a load applied at its end. It is assumed that one tooth sustains the entire load; *i. e.*, that there is contact only between one pair of teeth. This would be nearly true for gears with low numbers of teeth ; but in high numbered gears the force would be distributed over several pairs, and hence they would have an excess of strength. It is also assumed that the tooth has the same thickness from the pitch circle to its root. This is also nearly true for low numbered gears, while high numbered gears would have excess of strength as a result of this assumption. In Fig. 56 let $P =$ force at the pitch surface of the gear to be designed, $b =$ width of face, $l =$ depth of space, and $d =$ thickness of the tooth at the pitch circle. From Mechanics of Materials it is known that the moment of flexure, $Pl, = \dfrac{SI}{c}$; in which S is the unit stress in the outer fibre; I is the moment of inertia of the cross section, $= \dfrac{bd^3}{12}$; and c is the distance from the neutral axis to the outer fibre, $= \dfrac{d}{2}$. Hence, $Pl = \dfrac{Sbd^2}{6}$; $S = \dfrac{6Pl}{bd^2}$. Assume a value for diametral pitch, find the corresponding values of b, d, and l from table on page 58, and substitute in the above equation. S now becomes known, and may be compared with the ultimate strength of the material of the gear, $= S_1$. If the factor of safety, $= \dfrac{S_1}{S}$, is a proper value, the assumed pitch is right. If not, another pitch may be assumed and checked as before.

MACHINE DESIGN.

TABLE I.— FOR USE IN DESIGNING GEARS.

Diametral Pitch	Circular Pitch	Thickness of Tooth on the Pitch Line d	Width of Face b	Safe Stress for Cast Iron Gear Factor of Safety=10	Safe Unit Stress for Cast Iron Gear Factor of Safety=10	Safe Stress for Cast Steel Gear Factor of Safety=6	Safe Unit Stress for Cast Steel Gears Factor of Safety=6
½	6.283	3.141	20	15250	763	61000	3033
¾	4.189	2.094	13	6590	507	26390	2030
1	3.141	1.571	9	3442	382	13770	1530
1¼	2.513	1.256	7½	2250	305	9000	1220
1½	2.094	1.047	6	1530	255	6120	1020
1¾	1.795	.897	5½	1200	218	4800	872
2	1.571	.785	4½	862	192	3450	767
2¼	1.396	.698	4	684	170	2738	682
2½	1.256	.628	3½	532	152	2130	610
2¾	1.142	.571	3	417	139	1668	556
3	1.047	.523	2½	318	127	1272	509
3½	.897	.449	2	216	108	864	437
4	.785	.393	1¾	168	94	672	383
5	.628	.314	1⅝	124	76	498	306
6	.523	.262	1½	96	63	384	253
7	.449	.224	1⅜	74	54	296	216
8	.393	.196	1¼	60	48	240	192
9	.349	.174	1⅛	48	42	191	170
10	.314	.157	1 1⁄16	41	38	163	153
12	.262	.131	15⁄16	30	32	120	128
14	.224	.112	13⁄16	22	27	88	109

55. The work of approximation may be avoided by the use of Table I. Column 1 gives values of diametral pitch. Column 2 gives values of circular pitch. Column 3 gives values of d. Column 4 gives values of width of face, $= b$, corresponding to good practice. If this value of b is accepted, the table may be used as follows: The maximum working force at the pitch surface of the gear is found as above. If this value is found in column 5, the value of diametral pitch horizontally opposite in column 1 may be used for a cast iron gear, with a factor of safety of 10. If the value is not found in column 5, take the next greater value, and use the corresponding pitch. This will slightly increase the factor of safety.

If the conditions of the design require some different value for b, divide the maximum working force at the pitch surface by the

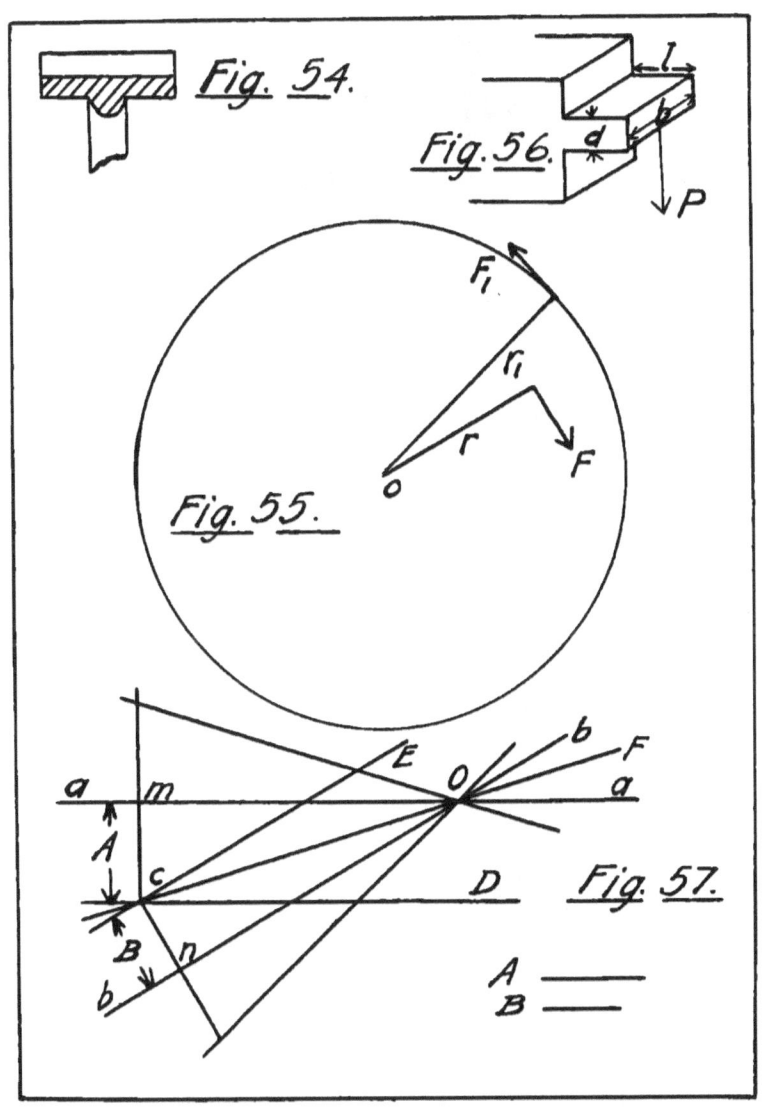

width of face, find this value, or the next greater, in column 6, and use the corresponding value of pitch.

If the pitch thus determined is too large for the design, a steel casting may be used, and the pitch will be determined by use of column 7 or 8 as above.

56. Non-Circular Wheels. — Only circular centrodes or pitch curves correspond to a constant velocity ratio; and by making the pitch curves of proper form, almost any variation in the velocity ratio may be produced. Thus a gear whose pitch curve is an ellipse, rotating about one of its foci, may engage with another elliptical gear, and if the driver has a constant angular velocity, the follower will have a constantly varying angular velocity. If the follower is rigidly attached to the crank of a slider crank chain, the slider will have a quick return motion. This is sometimes used for shapers and slotting machines. When more than one fluctuation of velocity per revolution is required, it may be obtained by means of "lobed gears"; *i. e.*, gears in which the curvature of the pitch curve is several times reversed. If a describing circle be rolled on these non-circular pitch curves, the tooth outlines will vary in different parts; hence, in order to cut such gears, many cutters would be required for each gear. Practically, this would be too expensive; and when such gears are used the pattern is accurately made, and the cast gears are used without "tooling" the tooth surfaces.

57. Bevel Gears. — All transverse sections of spur gears are the same, and their axes intersect at infinity. Spur gears serve to transmit motion between parallel shafts. It is necessary also to transmit motion between shafts whose axes intersect. In this case the pitch cylinders become pitch cones; the teeth are formed upon these conical surfaces, the resulting gears being called bevel gears. To illustrate, let a and b, Fig. 57, be the axes between which the motion is to be transmitted with a given velocity ratio. This ratio is equal to the ratio of the length of the line A to that of B. Draw a line CD parallel to a, at a distance from it equal to the length of the line A. Also draw the line CE parallel to b, at a distance from it equal to the length of the line B. Join the point of intersection of these lines to the point C, the intersection of the given axes. This locates the line CF, which is the line of contact of two pitch cones which will roll together to transmit the required velocity ratio.

For $\dfrac{mc}{nc} = \dfrac{A}{B}$, and if it be supposed that there are frusta of cones so thin that they may be considered cylinders, their radii being equal to mc and nc, it follows that they would roll together, if slipping be prevented, to transmit the required velocity ratio. But all pairs of radii of these pitch cones have the same ratio, $= \dfrac{mc}{nc}$, and therefore any pair of frusta of the pitch cones may be used to roll together for the transmission of the required velocity ratio. To insure this result, slipping must be prevented, and hence teeth are formed upon the selected frusta of the pitch cones. The theoretical determination of these may be explained as follows :

1st. *Cycloidal Teeth.*— If a cone A (Fig. 58), be rolled upon another cone, B, an element bc of the cone A will generate a conical surface, and a spherical section of this surface, adb, is called a spherical epicycloid. Also if a cone, A (Fig. 59), roll on the inside of another cone, C, an element bc of A will generate a conical surface, a spherical section of which, bda, is called a spherical hypocycloid. If now the three cones, B, C, and A, roll together, always tangent to each other on one line, as the cylinders were in the case of spur gears, there will be two conical surfaces generated by an element of A ; one upon the cone B, and another upon the cone C. These may be used for tooth surfaces to transmit the required constant velocity ratio. Because, since the line of contact of the cones is the axo* of the relative motion of the cones, it follows that a plane normal to the motion of the describing element of the generating cone at any time, will pass through this axo. And also, since the describing element is always the line of contact between

---

* An axo is an instantaneous axis, of which a centro is a projection.

the generated tooth surfaces, the normal plane to the line of contact of the tooth surfaces always passes through the axo, and the condition of rotation with a constant velocity ratio is fulfilled.

2d. *Involute Teeth.*—If two pitch cones are in contact along an element, a plane may be passed through this element, making an angle (say 75°) with the plane of the axes of the cones. Tangent to this plane there may be two cones, whose axes coincide with the axes of the pitch cones. If a plane be supposed to wind off from one base cone upon the other, the line of tangency of the plane with one cone will leave the cone and advance in the plane toward the other cone, and will generate simultaneously upon the pitch cones conical surfaces, and spherical sections of these surfaces will be spherical involutes. These surfaces may be used for tooth surfaces, and will transmit the required constant velocity ratio, because the tangent plane is the constant normal to the tooth surfaces at their line of contact, and this plane passes through the axo of the pitch cones.

To determine the tooth surfaces with perfect accuracy, it would be necessary to draw the required curves on a spherical surface, and then to join all points of these curves to the point of intersection of the axes of the pitch cones. Practically this would be impossible, and an approximate method is used.

If the frusta of pitch cones be given, B and C, Fig. 60, then points in the base circles of the cones, as L, M, and K, will move always in the surface of a sphere whose projection is the circle $LAKM$. Properly, the tooth curves should be laid out on the surface of this sphere, and joined to the centre of the sphere to generate the tooth surfaces. Draw cones LGM and MHK tangent to the sphere on circles represented in projection by lines LM and MK. If now tooth curves be drawn on these cones, with the base circle of the cones as pitch circles, they will very closely approximate the tooth curves that should be drawn on the spherical surface. But a cone may be cut along one of its elements and rolled out, or developed, upon a plane. Let MDH be a part of the cone MHK, developed, and let MNG be a part of the cone MGL, developed. The circular arcs MD and MN may be used just as pitch circles are in the case of spur gears, and the teeth may be laid out in exactly the same way, the curves being either cycloidal or involute, as required. Then the developed cones may be wrapped back, and the curves drawn may serve as directrices for the tooth surfaces, all of whose elements converge to the centre of the sphere of motion.

58. The teeth of spur gears may be cut by means of milling cutters, because all transverse sections are alike; but with bevel gears the conditions are different. The tooth surfaces are conical surfaces, and therefore the curvature varies constantly from one end of the tooth to the other. Also the thickness of the tooth and the width of space vary constantly from one end to the other. But the curvature and thickness of a milling cutter cannot vary, and therefore a milling cutter cannot cut an accurate bevel gear. Small bevel gears are, however, cut with milling cutters with sufficient accuracy for practical purposes. The cutter is made as thick as the narrowest part of the space between the teeth, and its curvature is made that of the middle of the tooth. Two cuts are made for each space. Let Fig. 61 represent a section of the cutter. For the first cut it is set relatively to the gear blank, so that the pitch point a of the cutter travels toward the apex of the pitch cone, and for the second cut so that the pitch point b travels toward the apex of the pitch cone. This method gives an approximation to the required form. Gears cut in this manner usually need to be filed slightly before they work satisfactorily. Bevel gears with absolutely correct tooth surfaces may be made by planing. Suppose a planer in which the tool point travels always in some line through the apex of the pitch cone. Then suppose that as it is slowly fed down the tooth surface, it is guided along the required tooth curve by means of a templet. From what has preceded it will be clear that the tooth so formed will be correct. Planers embodying these principles have been designed and constructed by Mr. Corliss of Providence, and Mr. Gleason of Rochester, with the most satisfactory results.

59. Design of Bevel Gears. — Given energy to be transmitted, rate

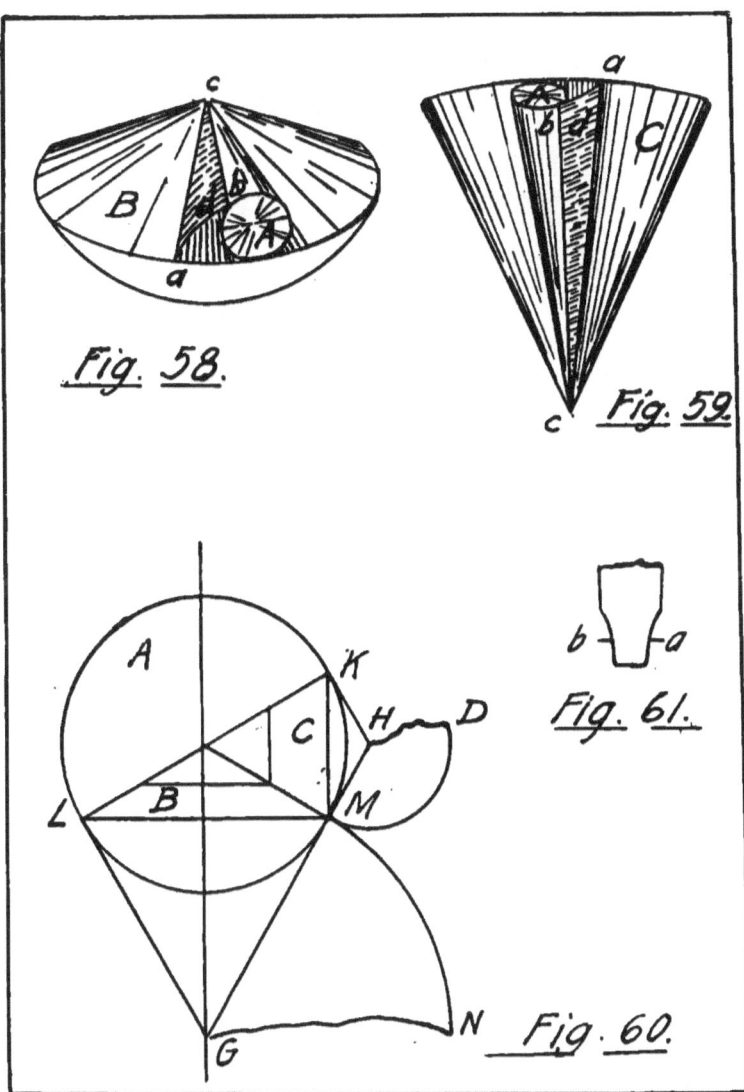

Fig. 58.

Fig. 59.

Fig. 61.

Fig. 60.

of rotation of one shaft, velocity ratio, and angle between axes ; to design a pair of bevel gears. Locate the intersection of axes, O, Fig. 62. Draw the axes OA and OB, making the required angle with each other. Locate OC, the line of tangency of the pitch cones, by the method given on page 59. Any pair of frusta of the pitch cones may be selected upon which to form the teeth. Special conditions of the problem usually dictate this selection approximately. Suppose that the inner limit of the teeth may be conveniently at D. Then make DP, the width of face, $= DO \div 2$. Or, if P is located by some limiting condition, lay off $PD = PO \div 3$. In either case the limits of the teeth are defined tentatively. Now from the energy and the number of revolutions of one shaft (either shaft may be used), the moment of torsion may be found. The mean force at the pitch surface = this torsional moment ÷ the mean radius of the gear ; $i.\ e.$, the radius of the point M, Fig. 62, midway between P and D. The pitch corresponding to this force may be found from Table I. This would be the mean pitch of the gears. But the pitch of bevel gears is measured at the large end, and diametral pitch varies inversely as the distance from O. In this case the distances of M and P from O are to each other as 5 is to 6. Hence the value of diametral pitch found from the table $\times \dfrac{5}{6} =$ the diametral pitch of the gear. If this value does not correspond with any of the usual values of diametral pitch, the next smaller value may be used. This would result in a slightly increased factor of safety. If the diametral pitch thus found, multiplied by the diameter corresponding to the point P, does not give an integer for the number of teeth, the point P may be moved outward along the line OC, until the number of teeth becomes an integer. This also would result in slight increase of the factor of safety. The point P is thus finally located, the corrected width of face $= PO \div 3$, and the pitch is known. The drawing of the gears may be completed as follows : Draw AB perpendicular to PO. With A and B as centres, draw the arcs PE and PF. Use these as pitch arcs, and draw the outlines of two or three teeth upon them, with cycloidal or involute curves as required. These will serve to show the form of tooth outlines. From P each way along the line AB lay off the addendum and the clearance. From the four points thus located draw lines toward O, terminating in the line DG. The tops of teeth and the bottoms of spaces are thus defined. Lay off upon AB below the bottoms of the spaces, a space about equal to the thickness of the tooth on the pitch circle. This gives a ring of metal to support the teeth. Join this to a properly proportioned hub as shown. The plan and elevation of each gear may now be drawn by the ordinary methods of projection.

60. Skew Bevel Gears. — Spur gears serve to communicate motion between parallel axes, and bevel gears between axes that intersect. But it is sometimes necessary to communicate motion between axes that are neither parallel nor intersecting. If the parallel axes are turned out of parallelism, or if intersecting axes are moved into different planes, so that they no longer intersect, the pitch surfaces become hyperbolic paraboloids in contact with each other along a straight line, which is the generatrix of the pitch surfaces. These hyperbolic paraboloids rolled upon each other, circumferential slipping being prevented, will transmit motion with a constant velocity ratio. There is, however, necessarily a slipping of the elements of the surfaces upon each other parallel to themselves. Teeth may be formed on these pitch surfaces, and they may be used for the transmission of motion between shafts that are not parallel nor in the same plane. Such gears are called "Skew Bevel Gears." The difficulties of construction and the additional friction due to slipping along the elements, make them undesirable in practice, and there is seldom a place where they cannot be replaced by some other form of connection.

A very complete discussion of the subject of Skew Bevel Gears may be found in Prof. McCord's "Kinematics."

61. Spiral Gearing. — If line contact is not essential there is much wider range of choice of gears to connect shafts which are neither parallel nor intersecting. A and B, Fig. 63, are axes of rotation in different planes, both planes being parallel to the paper. Let EF

and GH be cylinders on these axes, tangent to each other at the point S. Any line may now be drawn through S either between A and B, or coinciding with either of them. This line, say DS, may be taken as the common tangent to helical or screw lines drawn on the cylinders EF and GH; or helical surfaces may be formed on both cylinders, DS being their common tangent at S. **Spiral Gears** are thus produced. Each one is a portion of a many-threaded screw. The contact in these gears is point contact; in practice the point of contact becomes a very limited area.

62. When the angle between the shafts is made equal to 90°, and one gear has only one or two threads, it becomes a special case of spiral gearing known as **Worm Gearing**. In this special case the gear with a single or double thread is called the *worm*, while the other gear, which is still a many-threaded screw, is called the *worm wheel*. If a section of a worm and worm wheel be made on a plane passing through the axis of the worm, and normal to the axis of the worm wheel, the form of the teeth will be the same as that of a rack and pinion; in fact the worm, if moved parallel to its axis, would transmit rotary motion to the worm wheel. From the consideration of racks and pinions it follows that if the involute system is used, the sides of the worm teeth will be straight lines. This simplifies the cutting of the worm, because a tool may be used capable of being sharpened without special methods. If the worm wheel were only a thin plate the teeth would be formed like those of a spur gear, of the same pitch and diameter. But since the worm wheel must have greater thickness, and since all other sections parallel to that through the axis of the worm, as CD and AB, Fig. 64, show a different form and location of tooth, it is necessary to make the teeth of the worm wheel different from those of a spur gear, if there is to be contact between the worm and worm wheel anywhere except in the plane EF, Fig. 64. This would seem to involve great difficulty, but it is accomplished in practice as follows: A duplicate of the worm is made of tool steel, and "flutes" are cut in it parallel to the axis, thus making it into a cutter, which is tempered. It is then mounted in a frame in the same relation to the

worm wheel that the worm is to have when they are finished and in position for working. The distance between centres, however, is somewhat greater, and is capable of being gradually reduced. Both are then rotated with the required velocity ratio by means of gearing properly arranged, and the cutter or "hob" is fed against the worm wheel till the distance between centres becomes the required value. The teeth of the worm wheel are "roughed out" before they are "hobbed." By the above method the worm is made to cut its own worm wheel.*

Fig. 65 represents the half section of a worm. If it is a single worm the thread A, in going once around, comes to B; twice around, to C; and so on. If it is a double worm the thread A, in going once around, comes to C, while there is an intermediate thread, B. It follows that if the single worm turns through one revolution it will push *one* tooth of the worm wheel with which it engages, past the line of centres; while the double worm will push *two* teeth of the worm wheel past the line of centres. The single worm, therefore, must make as many revolutions as there are teeth in the worm wheel, in order to cause one revolution of the worm wheel; while for the same result the double worm only needs to make half as many revolutions. The ratio of the angular velocity of a single worm to that of the worm wheel with which it engages is $=\frac{n}{1}$, in which n equals the number of teeth in the worm wheel. For the double worm this ratio is $\frac{n}{2}$.

Worm gearing is particularly well adapted for use where it is necessary to get a high velocity ratio in limited space.

The pitch of a worm is measured parallel to the axis of rotation. The pitch of a single worm is p, Fig. 65. It is equal to the circular pitch of the worm wheel. The pitch of a double worm is $p_1 = 2p = 2 \times$ circular pitch of the worm wheel.

* This subject is fully treated in Unwin's "Elements of Machine Design," and in Brown & Sharpe's "Treatise on Gearing."

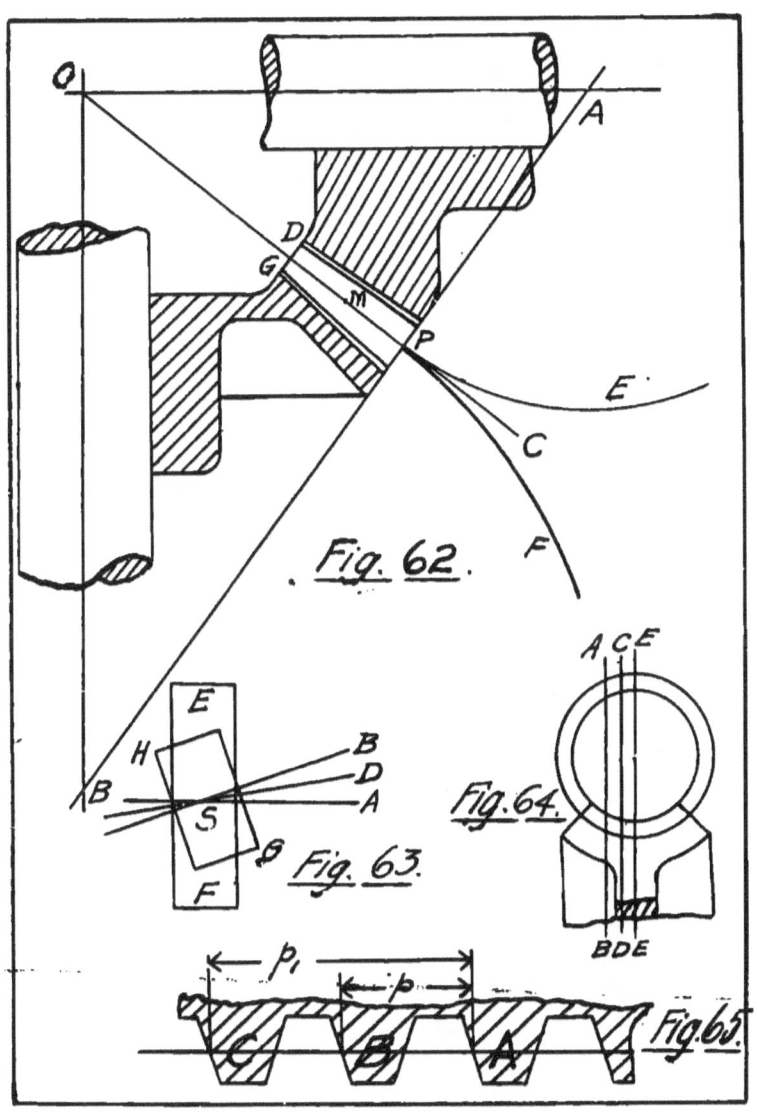

Fig. 62.

Fig. 63.

Fig. 64.

Fig. 65.

TOOTHED WHEELS, OR GEARS. 67

63. Design of Worm Gears. — Let $E =$ energy to be transmitted through the worm wheel per minute; $N =$ number of revolutions per minute; $R =$ pitch radius of the worm wheel; $F =$ force at pitch surface of the worm wheel. Then

$$E = 2\pi RN \times F = \text{space factor} \times \text{force factor},$$

and $$F = \frac{E}{2\pi RN} = \frac{\text{torsional moment}}{R}.$$

Hence, if E, R, and N are given, F becomes known, and an approximate value for pitch may be found from Table I. If p is the diametral pitch thus found, the corresponding circular pitch $= \frac{\pi}{p}$ $=$ pitch of a single worm to mesh with the worm wheel.* The pitch of a double worm to mesh with the worm wheel $= \frac{2\pi}{p}$. The number of teeth, n, in the worm wheel $= 2Rp$. For a single worm the velocity ratio $= \frac{n}{1}$; for double worm $= \frac{n}{2}$. The rate of rotation of the single worm $= Nn$; of the double worm $= \frac{Nn}{2}$. From the energy, E_1, to be transmitted per minute by the worm shaft, and the rate of rotation, N_1, the moment of torsion is found $= \frac{E_1}{2\pi N_1}$. From this a suitable belt driving mechanism may be designed by methods to be given later.

64. When the worm and worm wheel are determined, a working drawing may be made as follows: Draw AB, Fig. 66, the axis of the worm wheel, and locate O, the projection of the axis of the worm, and P, the pitch point. With O as a centre draw the pitch, full depth, and addendum circles, G, H, and K; also the arcs CD and EF, bounding the tops of the teeth and the bottoms of the spaces of the worm wheel. Make the angle $\beta = 90°$. Below EF lay off a proper thickness of metal to support the teeth, and join this by the

* This value must be made such that it may be cut in an ordinary lathe. See next page.

web LM to the hub N. The tooth outlines in the other sectional view are drawn exactly as for an involute rack and pinion. Full views might be drawn, but they involve difficulties of construction, and do not give any additional information to the workman.

65. Solution from Other Data.—Two shafts, at right angles to each other in different planes, are to be connected by means of worm gearing. The maximum distance between them is fixed, and its value, d, is given. The required velocity ratio, r, the rate of rotation of the worm shaft, N, and the energy to be transmitted per minute, E, are also given. It is required to design the gears. A single thread worm is to be used. Let $E = 33000$ ft.lbs. per minute; $r = 40$, i. e., the worm makes 40 revolutions per revolution of the worm wheel; $d = 8"$; $N = 280$ revolutions per minute. The velocity ratio depends upon the pitch of the worm, but not upon its diameter. Because, whatever the diameter of the worm, it pushes one tooth of the worm wheel past the line of centres (QR, Fig. 66) each revolution. The pitch diameter of the worm may therefore be any convenient value. The worm wheel must have 40 teeth in order that the single thread worm shall turn 40 times to turn it once. The number of threads per inch of the worm, measured axially, is the reciprocal of the pitch. This should be such a value that it may be cut in an ordinary lathe without special arrangement of the change gears. Lathes of medium size are capable of cutting 1, 2, 3, 4, etc., threads per inch. The circular pitch may therefore be 1, 0·5, 0·334, 0·25, etc. In the case considered suppose that the pitch diameter of the worm may conveniently be about 1·5". The corresponding pitch radius of the worm wheel $= 8" - 1·5" = 6·5"$. Since there must be 40 teeth, it follows that the circular pitch, $= \dfrac{2\pi 6 \cdot 5}{40} = 1 \cdot 021$. This should be exactly 1, as indicated above. Let circular pitch $= 1$, and check for strength. The tooth may be considered as a cantilever, as in the case of spur gear teeth. Then $S = \dfrac{6Pl}{bd^2}$, in which $P =$ force at the pitch surface, $l =$ depth of space, $b =$ distance corresponding to the arc EF, Fig. 66, and

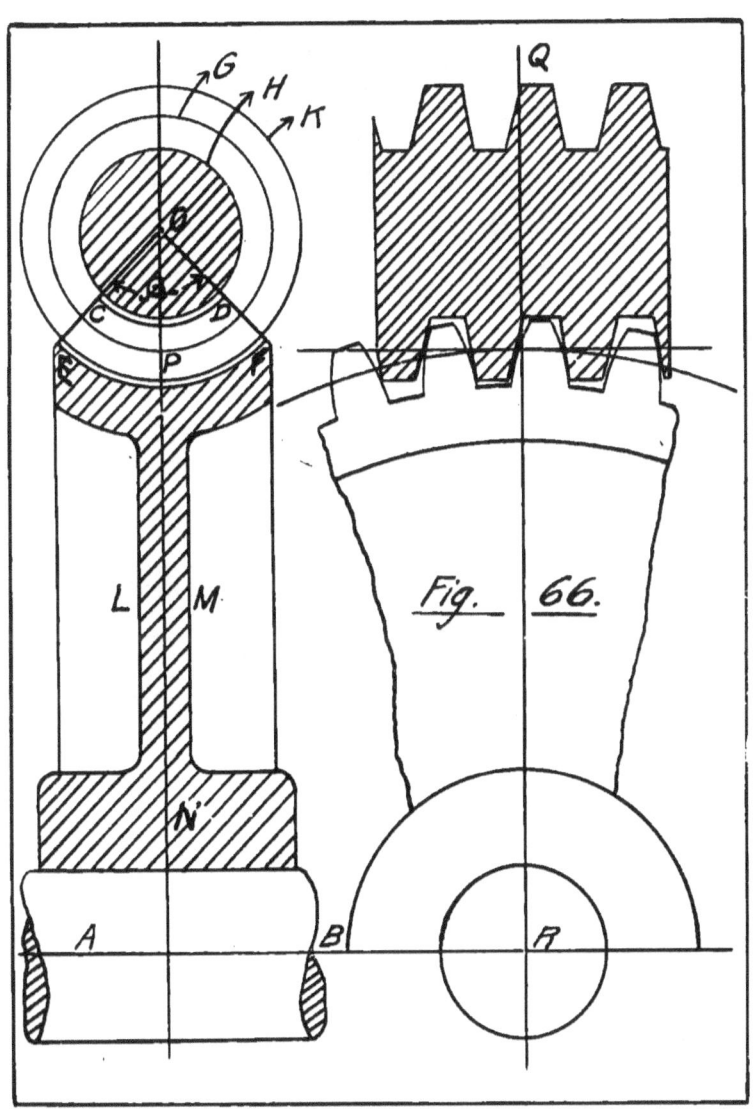

Fig. 66.

TOOTHED WHEELS, OR GEARS. 69

$d =$ thickness of tooth at pitch circle $= \frac{1}{2}$ circular pitch. — To find P. The energy $= 33,000$ ft.lbs. per minute ; $N =$ rate of rotation of the worm wheel shaft $= \frac{280}{40} = 7$; the pitch radius of the worm wheel corrected for 1 pitch $= \frac{40}{2\pi} = 6\cdot 37'' = 0\cdot 53$ feet ; the torsional moment at the worm wheel shaft $= \frac{33000}{2\pi N} = Pr$. Hence,

$$P = \frac{33000}{2\pi r N} = \frac{33000}{2\pi \times 0\cdot 53 \times 7} = 1414 \text{ lbs.}; \quad l = 0\cdot 64''; \quad d = 0\cdot 5''.$$

To find b. The pitch radius of the worm corrected for 1 pitch $= 8 - 6\cdot 37 = 1\cdot 63$. The radius of the outside of the worm $= 1\cdot 63 +$ addendum, $0\cdot 318, = 1\cdot 948$; the arc subtended by $90°$ on this radius $= 90° \times 0\cdot 0174 \times 1\cdot 948 = 3\cdot 05 = b$. Hence $S = \frac{6 \times 1414 \times 0\cdot 64}{3\cdot 05 \times 0\cdot 25} = 7121$ lbs. If cast iron were to be used, this would give a factor of safety $= \frac{20000}{7121} = 2\cdot 8$. This is too small, and a larger circular pitch would need to be used, and the worm would have to be cut in a lathe capable of cutting less than 1 thread per inch. If steel casting is an allowable material the factor of safety would $= \frac{50000}{7121} = 7 +$. This is a proper value, and the design is correct.

This method of design applies when the worm wheel is cut by a "hob."

66. Compound Spur Gear Chains. — Spur gear chains may be compound, *i. e.*, they may contain links which carry more than two elements. Thus in Fig. 67 the links a and d each carry three elements. In the latter case the teeth of d must be counted as two elements, because by means of them d is paired with both b and c. In the case of the three-link spur gear chain the wheels b and c meshed with each other, and a point in the pitch circle of c moved with the same linear velocity as a point in the pitch circle of b, but in the opposite direction. In Fig. 67 points in all the pitch circles have the same linear velocity, since the motion is equivalent to rolling together of the pitch circles without slipping ; but c and b now rotate in the same direction. Hence it is seen that the introduction of the wheel d has reversed the direction of rotation, without changing the velocity ratio. The size of the wheel d, which is called an "idler," has no effect upon the motion of c and b. It simply receives, upon its pitch circle, a certain linear velocity from c, and transmits it unchanged to b. Hence the insertion of any number of idlers does not affect the velocity ratio of c to b, but each added idler reverses the direction of the motion. Thus, with an odd number of idlers, c and b will rotate in the same direction ; and with an even number of idlers, c and b will rotate in opposite directions.

If parallel lines be drawn through the centres of rotation of a pair of gears, and if from the centres distances be laid off on these lines inversely proportional to the angular velocities of the gears, then a line joining the points so determined will cut the line of centres in a point which is the centro of the gears. In Fig. 67, since the rotation is in the same direction, the lines have to be laid off on the same side of the line of centres. The pitch radii are inversely proportional to the angular velocities of the gears, and hence it is only necessary to draw a tangent to the pitch circles of b and c, and the intersection of this line with the line of centres is the centro, bc, of c and b. The centroids of c and b are c_1 and b_1, circles through the point bc, about the centres of c and b. Obviously, this four link mechanism may be replaced by a three link mechanism, in which c is an annular wheel meshing with a pinion b. The four link mechanism is more compact, however, and usually more convenient in practice.

67. The other principal form of spur gear chain is shown in Fig. 68. The wheel d has two sets of teeth of different pitch diameter, one pairing with c, and the other with b. The point bd now has a different linear velocity from cd, greater or less in proportion to the ratio of the radii of those points. The angular velocity ratio may be obtained as follows :

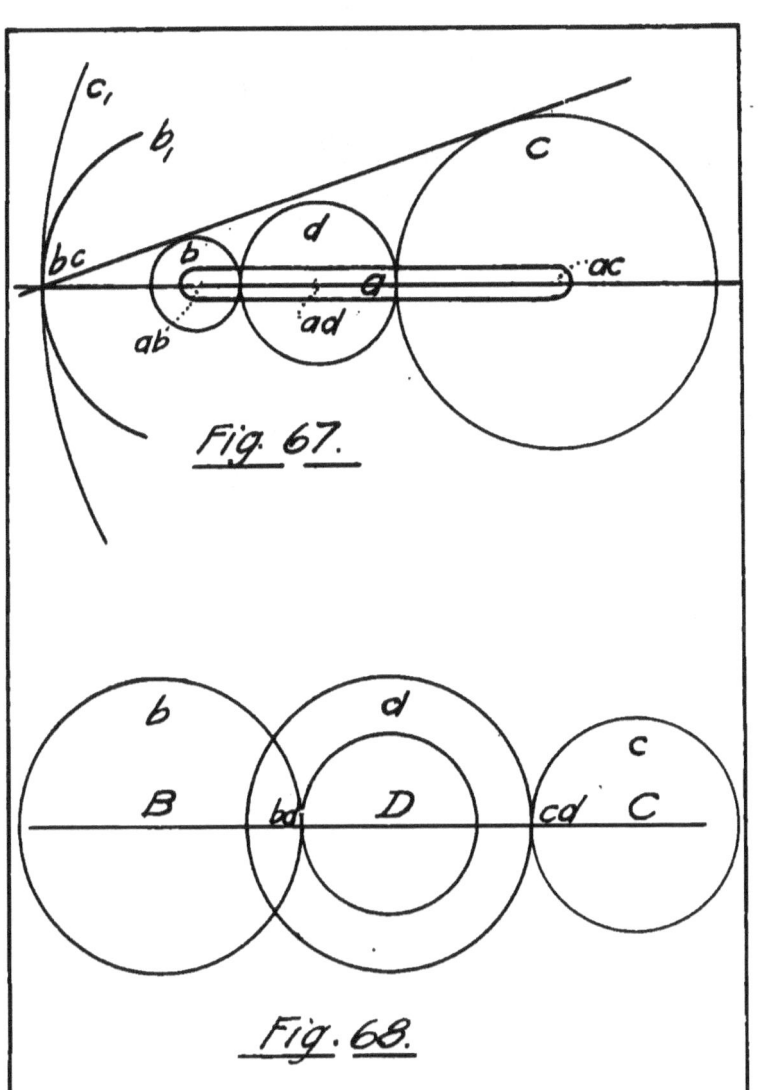

Fig. 67.

Fig. 68.

TOOTHED WHEELS, OR GEARS.

$$\frac{\text{angular veloc. } d}{\text{angular veloc. } c} = \frac{Ccd}{Dcd};$$

also

$$\frac{\text{angular veloc. } b}{\text{angular veloc. } d} = \frac{Dbd}{Bbd}.$$

Multiplying,

$$\frac{\text{angular veloc. } b}{\text{angular veloc. } c} = \frac{Ccd \times Dbd}{Dcd \times Bbd}.$$

The numerator of the last term consists of the product of the radii of the "followers"; and the denominator consists of the product of the radii of the "drivers." The diameters or numbers of teeth could be substituted for the radii.

In general, the angular velocity of the first driver is to the angular velocity of the last follower as the product of the number of teeth of the followers is to the product of the number of teeth of the drivers. This applies equally well to compound spur gear trains that have more than three axes. Therefore, in *any* spur gear chain the velocity ratio equals the product of the number of teeth in the followers divided by the product of the number of teeth in the drivers. The direction of rotation is reversed if the number of intermediate axes is even, and is not reversed if the number is odd. If the train includes annular gears their axes would be omitted from the number, because annular gears do not reverse the direction of rotation.

CHAPTER VI.

CAMS.

68. A machine part of irregular outline, as A, Fig. 69, may rotate or vibrate about an axis O, and communicate motion by line contact to another machine part, B. A is called a cam. A cylinder A, Fig. 70, having a groove of any form in its surface, may rotate about its axis, CD, and communicate motion to another machine part, B. A is a cam. A disc A, Fig. 71, having a groove in its face, may rotate about its axis, O, and communicate motion to another machine part, B. A is a cam. In fact it is only a modification of A, Fig. 69. In designing cams it is customary to consider a number of simultaneous positions of the driver and follower. The cam curve can usually be drawn from data thus obtained.

69. *Case I.*—The follower is guided in a straight line, and the contact of the cam with the follower is always in this line. The line may be in any position relatively to the centre of rotation of the cam; hence it is a general case. The point of the follower which bears on the cam is constrained to move in the line MN, Fig. 72. O is the centre of rotation of the cam. About O, as a centre, draw a circle tangent to MN at J. Then A, B, C, etc., are points in the cam. When the point A is at J the point of the follower which bears on the cam must be at A'; when B is at J the follower point must be at B'; and so on through an entire revolution. Through A, B, C, etc., draw lines tangent to the circle. With O as a centre, and OA' as a radius, draw a circular arc $A'A''$, intersecting the tangent through A at A''. Then A'' will be a point in the cam curve. For, if A returns to J, AA'' will coincide with JA', A'' will coincide with A', and the cam will hold the follower in the required position.

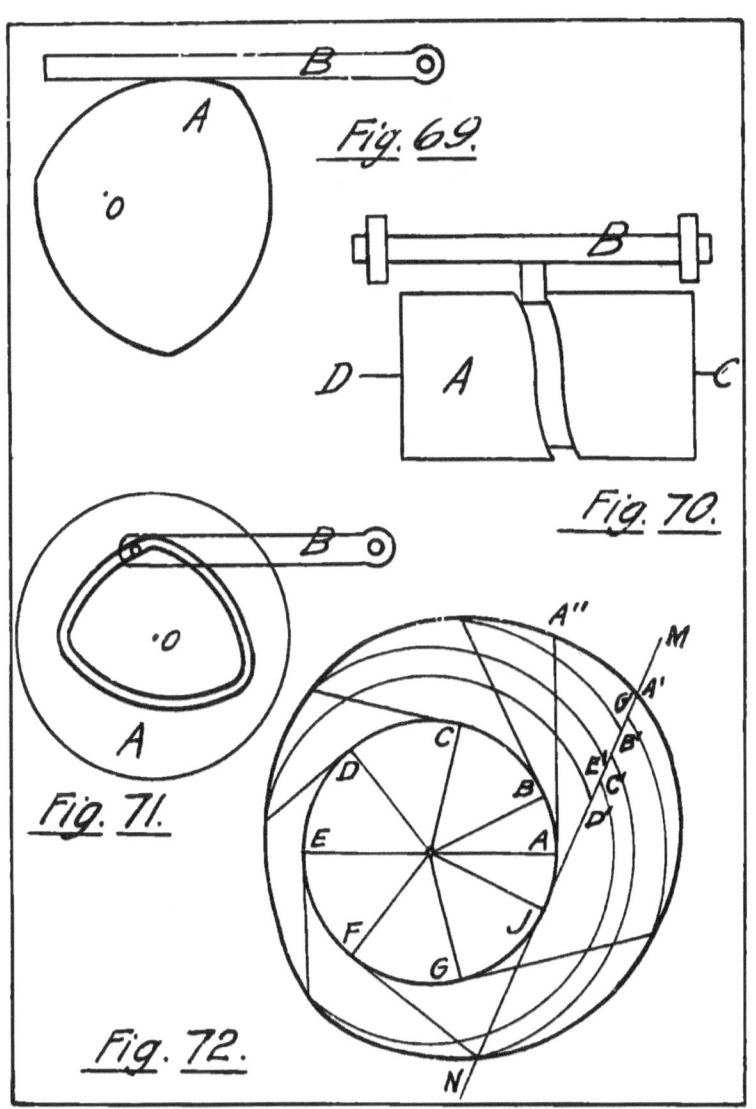

The same process for the other positions locates other points of the cam curve. A smooth curve drawn through these points is the required cam outline. Often, to reduce friction, a roller attached to the follower rests on the cam, motion being communicated through it. The curve found as above will be the path of the axis of the roller. The cam outline will then be a curve drawn inside of, and parallel to, the path of the axis of the roller, at a distance from it equal to the roller's radius. Contact between the follower and the cam is not confined to the line MN if a roller is used.

70. *Case II.*— The cam engages with a surface of the follower, and this surface is guided so that all of its positions are parallel. The method given is due to Professor J. H. Barr. O, Fig 73, is the centre of rotation of the cam. The follower surface occupies the successive positions 1, 2, 3, etc., when the lines A, B, C, etc. of the cam coincide with the vertical line through C. It is required to draw the outline of a cam to produce the motion required. Produce the vertical line through O, cutting the positions of the follower surface in A', B', C', etc. With O as a centre and radii OB', OC', etc., draw arcs cutting the lines B, C, D, etc. in the points B'', C'', D'', etc. Position 1 is the lowest position of the follower surface ; therefore A must be in contact with the follower surface in the vertical line through O, because if the tangency be at any other point the motion in one direction or the other will lower the follower, which is not allowable. A is therefore one point in the cam curve. Draw a line MN through B'' at right angles to $B''O$, and rotate $B''O$ till it coincides with $B'O$. Then the line MN will coincide with the position of the follower surface $2B'$. But the cam curve must be tangent to this line when B coincides with $B'O$, and therefore the line MN is a line to which the cam curve must be tangent. Similar lines may be drawn through the points C'', D'', etc. Each will be a line to which the cam curve must be tangent. Therefore, if a smooth curve be drawn tangent to all these lines, it will be the required cam outline.

71. *Case III.* — This is the same as Case II, except that the positions of the follower surface instead of being parallel, converge to a point, O, Fig. 74, about which the follower vibrates. The solution is the same as in Fig. 73, except that the angle between the lines corresponding to MN, Fig. 73, and the radial lines, instead of being a right angle, equals the angle between the corresponding position of the follower surface and the vertical.

In these cases the cam drives the follower in only one direction; the force of gravity, the expansive force of a spring, or some other force must hold it in contact with the cam. To drive the follower in both directions, the cam surface must be double, *i.e.*, it takes the form of a groove engaging with a pin or roller attached to the follower, as in Fig. 71. The foregoing principles apply to the laying out of the curves.

72. Case IV.—To lay out a cam groove on the surface of a cylinder.— A, Fig. 75, is a cylinder which is to rotate continuously about its axis. B can only move parallel to the axis of A. B may have a projecting roller to engage with a groove in the surface of A. CD is the axis of the roller in its mid-position. EF is the development of the surface of the cylinder. During the first quarter revolution of A, CD is required to move one inch toward the right with a constant velocity. Lay off $GH = 1''$, and $HJ = \frac{1}{4}KF$, locating J. Draw GJ, which will be the middle line of the cam groove. During the next half revolution of A, the roller is required to move two inches toward the left with a uniformly accelerated velocity. Lay off $JL = 2''$, and $LM = \frac{1}{2}KF$. Divide LM into any number of equal parts, say 4. Divide JL into 4 parts, so that each is greater than the preceding one by an equal increment. This may be done as follows : $1 + 2 + 3 + 4 = 10$. Lay off from J, $0.1\,JL$, locating a; then $0.2\,JL$ from a, locating b; and so on. Through a, b, and c draw vertical lines; through m, n, and o draw horizontal lines. The intersections locate d, e, and f. Through these points draw the curve from J to M, which will be the required middle line of the cam groove. During the remaining quarter revolution the roller is required to return to its starting point with a uniformly accelerated velocity. The curve MN is drawn in the same way as JM. On each side of the line $GJMN$ lay off parallel lines, their distance apart being equal to the diameter of the roller. Wrap EF upon the cylinder, and the required cam groove is located.

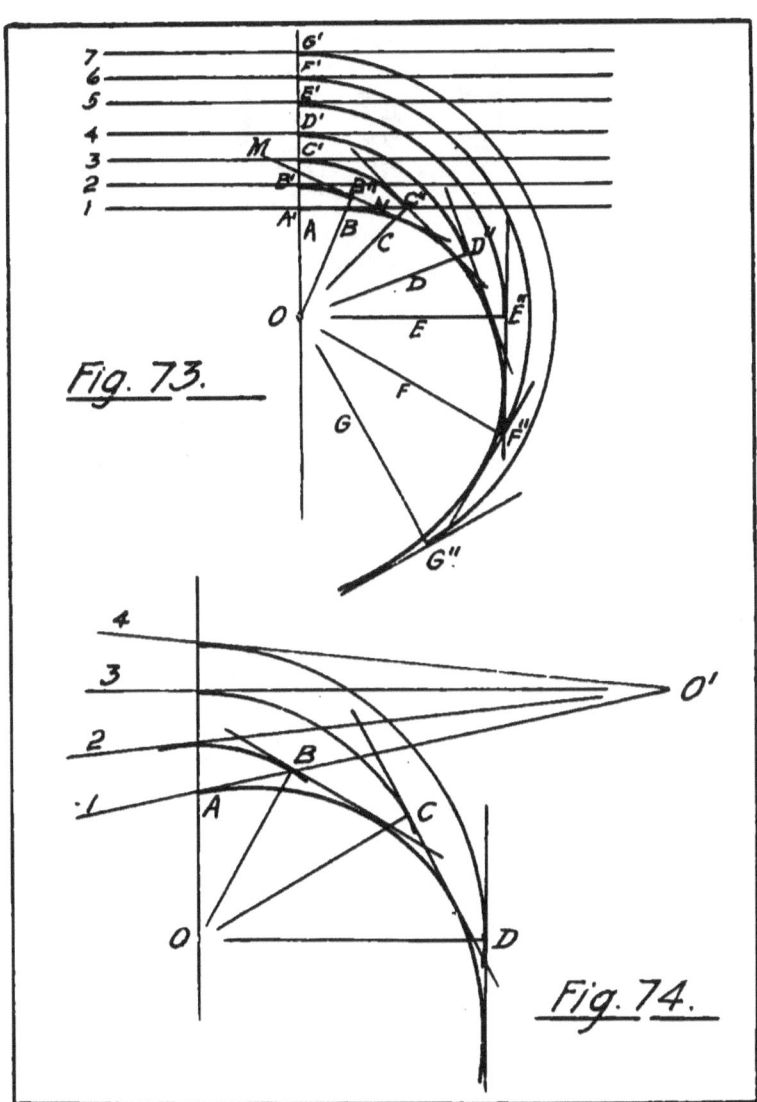

Fig. 73.

Fig. 74.

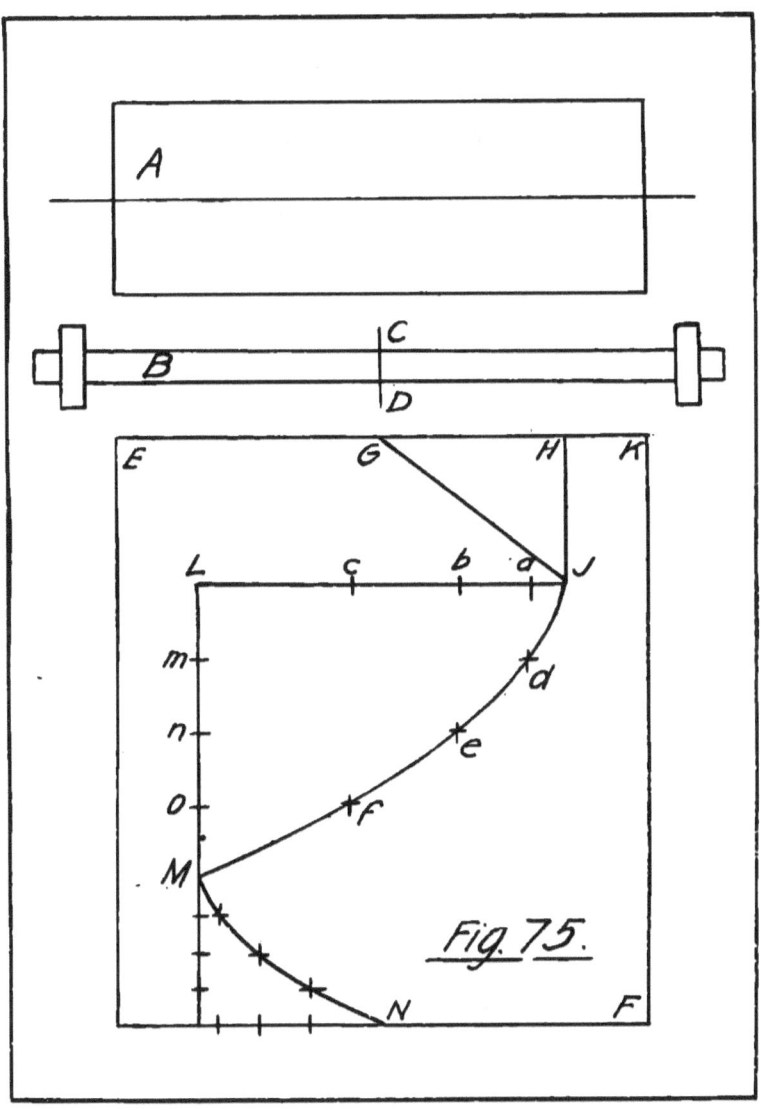

Fig. 75.

CHAPTER VII.

BELTS.

73. Transmission of Motion by Belts. — In Fig. 76, let A and B be two cylindrical surfaces, free to rotate about their axes; let CD be their common tangent, and let CD represent an inextensible connection between the two cylinders. Since it is inextensible, the points D and C, and hence the surfaces of the cylinders, must have the same linear velocity. Two points having the same linear velocity, and different radii, have angular velocities which are inversely proportional to their radii. Hence, since the surfaces of the cylinders have the same linear velocity, their angular velocities are inversely proportional to their radii. This is true of all cylinders connected by inextensible connectors. Suppose the cylinders to become pulleys, and the tangent line to become a belt. Let $C'D'$ be drawn; this becomes a part of the belt, making it endless, and rotation may be continuous. The belt will remain always tangent to the pulleys, and will transmit such rotation that the angular velocity ratio will constantly be the inverse ratio of the radii of the pulleys.

The case considered corresponds to a crossed belt, but the same reasoning applies to an open belt. See Fig. 77. A and B are two pulleys, and $CDD'C'C$ is an open belt. Since the points C and D are connected by a belt that is practically inextensible, the linear velocity of C and D is the same; therefore the angular velocities of the pulleys are to each other inversely as their radii. If the pulleys in either case were pitch cylinders of gears the conditions of velocity would be the same. In the first case, however, the direction of motion is reversed, while in the second case it is not. Hence the first corresponds to gears meshing directly with each other, while the second corresponds to the case of gears connected by an idler, or to the case of an annular gear and pinion.

Of course it is necessary that a belt should have some thickness; and, since the centre of pull is the centre of the belt, it is necessary to add to the radius of the pulley half of the thickness of the belt. The motion communicated by means of belting, however, does not need to be absolutely correct, and therefore in practice it is usually customary to neglect the thickness of the belt. The proportioning of pulleys for the transmission of any required velocity ratio is now a very simple matter.

Illustration. — A line shaft runs 150 revolutions per minute, and is supported by hangers with 16" "drop." It is required to transmit motion from this shaft to a dynamo to run 1800 revolutions per minute. A 30" pulley is the largest that can be conveniently used with 16" hangers. Let x = the diameter of required pulley for the dynamo; then from what has preceded $x \div 30 = 150 \div 1800$, and therefore $x = 2.5"$. But a pulley less than 4" diameter should not be used on a dynamo. Suppose in this case that it is 6". It is then impossible to obtain the required velocity ratio with one change of speed, *i. e.*, with one belt. Two changes of speed may be obtained by the introduction of a counter shaft. By this means the velocity ratio is divided into two factors. If it is wished to have the same change of speed from the line shaft to the counter as from the counter to the dynamo, then each velocity ratio would be $\sqrt{(1800 \div 150)} = \sqrt{12} = 3.46$. But this gives an inconvenient fraction, and the factors do not need to be equal. Let the factors be 3 and 4. See Fig. 78. A represents the line shaft, B the counter, and C the dynamo shaft. The pulley on the line shaft is 30", and the speed is to be three times as great at the counter, and therefore the pulley must have a diameter one-third as great, $= 10"$. The pulley on the dynamo is 6" diameter and the counter shaft is to run one-fourth as fast, and therefore the pulley on the counter opposite the dynamo pulley must be four times as large as the dynamo pulley, $= 24"$.

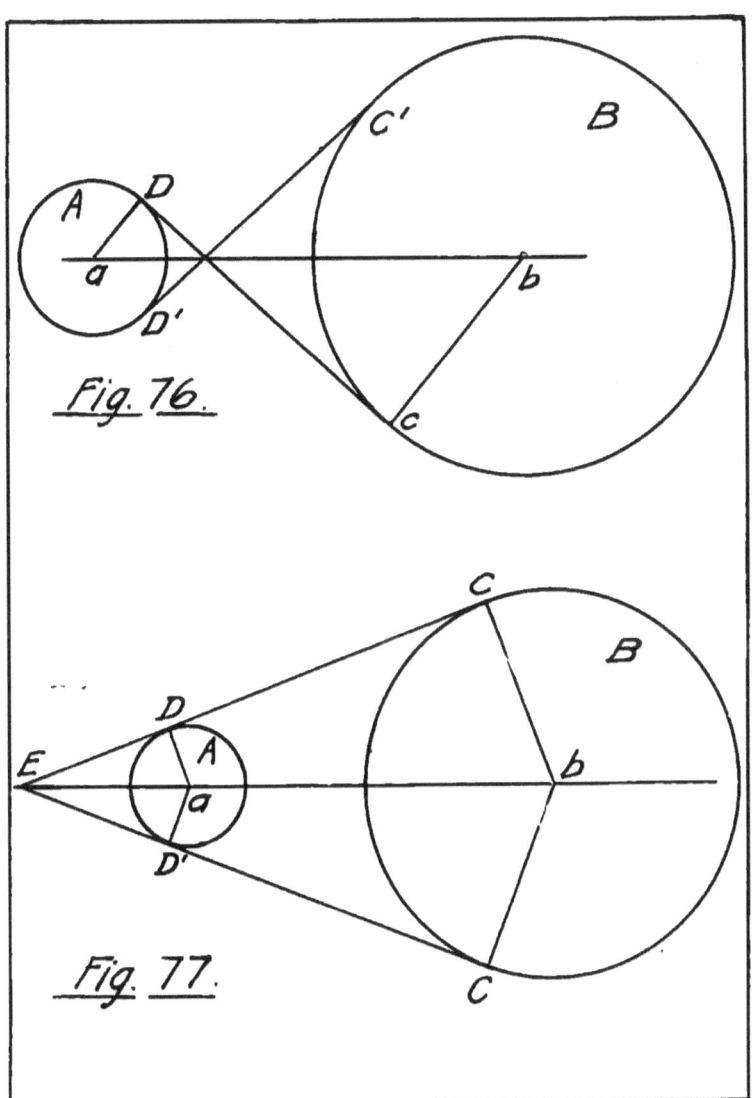

Fig. 76.

Fig. 77.

74. A belt may be shifted from one part of a pulley to another by means of pressure against the side which advances toward the pulley. Thus, if in Fig. 79 the rotation be as indicated by the arrow, and side pressure be applied at A, the belt will be pushed to one side, as is shown, and will consequently be carried into some new position on a pulley further to the left as it advances. Hence, in order that a belt may maintain its position on a pulley, *the centre line of the advancing side of the belt must be perpendicular to the axis of rotation.* When this condition is fulfilled the belt will run and transmit the required motion, regardless of the relative position of the shafts.

75. In Fig. 80, the axes AB and CD are parallel to each other, the above stated condition is fulfilled, and the belt will run correctly; but if the axis CD were turned into some new position, as $C'D'$, the side of the belt that advances toward the pulley E, cannot have its centre line in a plane perpendicular to the axis, and therefore it will run off. But if a plane be passed through the line CD, perpendicular to the plane of the paper, then the axis may be swung in this plane in such a way that the necessary condition shall be fulfilled, and the belt will run properly. This gives what is known as a "twist" belt, and when the angle between the shafts becomes 90°, it is a "quarter twist" belt. To make this clearer, see Fig. 81. Rotation is transmitted from A to B by an open belt, and it is required to turn the axis of B out of parallelism with that of A. The direction of rotation is as indicated by the arrows. Draw the line CD. If now the line CD be supposed to pass through the centre of the belt at C and D, it may become an axis, and the pulley B and the part of the belt FC may be turned about it, while the pulley A and the part of the belt ED remain stationary. During this motion the centre line of the part of the belt CF, which is the part that advances toward the pulley B when rotation occurs, is always in a plane perpendicular to the axis of the pulley B. The part ED, since it has not been moved, has also its centre line in a plane perpendicular to the axis of A. Therefore, the pulley B may be swung into any angular position about CD as an axis, and the condition of proper belt transmission will not be interfered with.

76. If the axes intersect, the motion can be transmitted between them by belting only by the use of "guide" or "idler" pulleys. Let AB and CD, Fig. 82, be intersecting axes, and let it be required to transmit motion from one to the other by means of a belt running on the pulleys E and F. Draw centre lines EK and FH through the pulleys. Draw the circle, G, of any convenient size, tangent to the lines EK and FH. In the axis of the circle G, let a shaft be placed on which are two pulleys, their diameters being equal to that of the circle G. These will serve as guide pulleys for the upper and lower sides of the belt, and by means of them the centre lines of the advancing parts of both sides of the belt will be kept in planes perpendicular to the axis of the pulley toward which they are advancing, the belts will run properly, and the motion will be transmitted as required.

77. An analogy will be noticed between gearing and belting for the transmission of rotary motion. Spur gearing corresponds to an open or crossed belt, transmitting motion between parallel shafts. Bevel gears correspond to a belt running on guide pulleys, transmitting motion between intersecting shafts. Skew bevel and spiral gears correspond to a "twist" belt, transmitting motion between shafts that are neither parallel nor intersecting.

78. If a flat belt be put on a "crowning" pulley, as in Fig. 83, the tension on AB will be greater than on CD, the belt will tend to be shifted into the position shown by the dotted lines E and F, and as rotation goes on, the belt will be carried toward the high part of the pulley, i.e., it will tend to run in the middle of the pulley. This is the reason why nearly all belt pulleys, except those on which the belt has to be shifted into different positons, are turned "crowning."

79. **Cone Pulleys.**— In performing different operations on a machine or the same operations on materials of different degrees of hardness, different speeds are required. The simplest way of obtaining them is by the use of cone pulleys. One pulley has a series of steps, and the opposing pulley has a corresponding series of steps. By shifting the belt from one pair to another the velocity

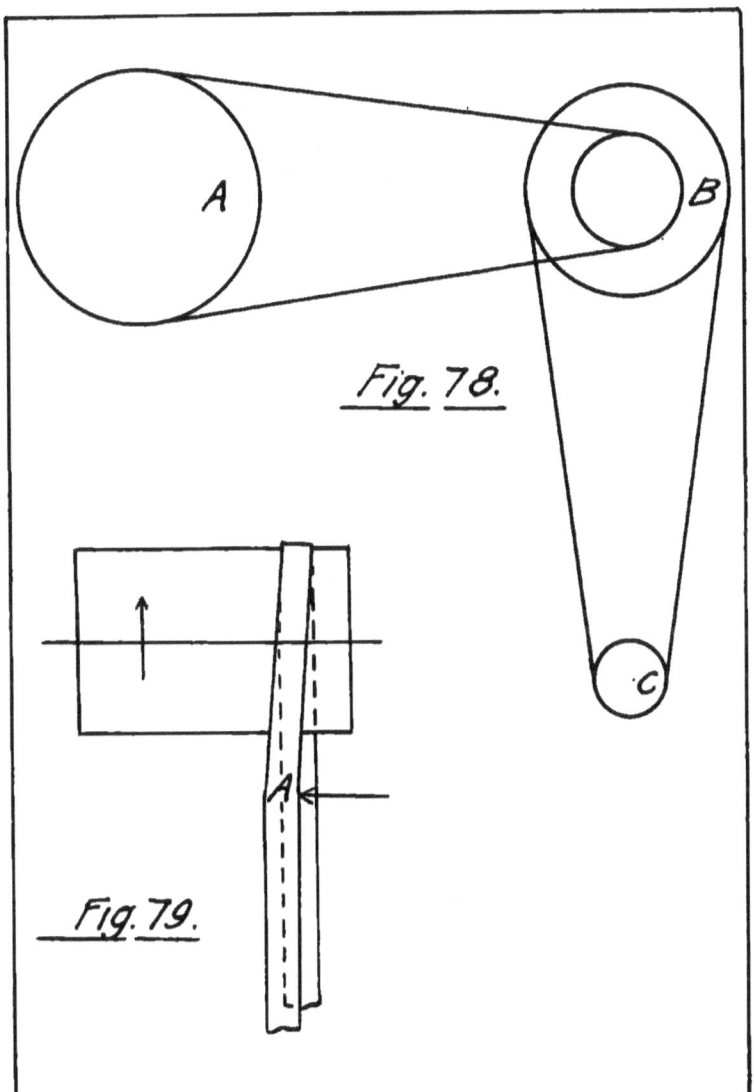

Fig. 78.

Fig. 79.

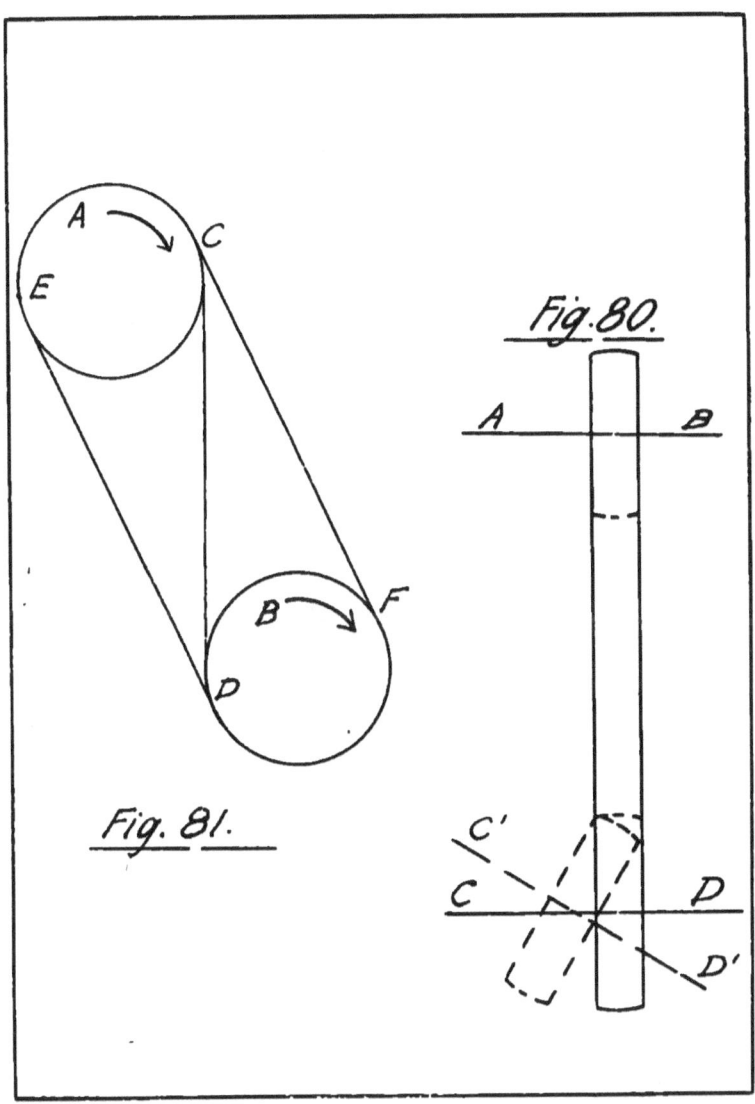

Fig. 80.

Fig. 81.

ratio is changed. Since the same belt is used on all the pairs of steps, they must be so proportioned that the belt length for all the pairs shall be the same; otherwise the belt would be too tight on some of the steps and too loose on others. Let the case of a crossed belt be first considered. The length of a crossed belt may be expressed by the following formula: Let $L =$ length of the belt; $d =$ distance between centres of rotation; $R =$ radius of the larger pulley; $r =$ radius of the smaller pulley. See Fig. 84. Then $L = 2\sqrt{d^2-(R+r)^2} + (R+r)(\pi + 2$ arc whose sine is $R + r \div d)$. In the case of a crossed belt, if the size of steps be changed so that the sum of their radii remains constant, the belt length will be constant. For in the formula the only variables are R and r, and these terms only appear in the formula as $R + r$; but $R + r$ is by hyothesis constant. Therefore any change that is made in the variables R and r, so long as their sum is constant, will not affect the value of the equation, and hence the belt length will be constant. It will now be easy to design cone pulleys for crossed belt. Suppose a pair of steps given to transmit a certain velocity ratio. It is required to find a pair of steps that will transmit some other velocity ratio, the length of belt being the same in both cases. Let r and $r' =$ radii of the given steps; R and $R' =$ radii of the required steps; $r + r' = R + R' = a$; the velocity ratio of R to $R' = b$. There are two equations between R and R', $R \div R' = b$, and $R + R' = a$. Combining and solving, it is found that $R' = a \div (1 + b)$, and $R = a - R'$. For an open belt the formula for length is: $L = 2\sqrt{d^2 - (R-r)^2} + \pi(R + r) + 2(R-r)$ (arc whose sine is $R - r \div d$). If R and r be changed as before, the term $R - r$ would of course not be constant, and two of the terms of the equation would vary in value; therefore the length of the belt would vary. The determination of cone steps for open belts therefore becomes a more difficult matter, and approximate methods are almost invariably used.

80. The following graphical approximate method is due to Mr. C. A. Smith, and is given, with full discussion of the subject, in "Transactions of the American Society of Mechanical Engineers,"

Vol. X, p. 269. Suppose first that the diameters of a pair of cone steps that transmit a certain velocity ratio are given, and that the diameters of another pair that shall serve to transmit some other velocity ratio are required. The distance between centres of axes is given. See Fig. 85. Locate the pulley centres O and O', at the given distance apart; about these centres draw circles whose diameters equal the diameters of the given pair of steps; draw a straight line GH, tangent to these circles; at J, the middle point of the line of centres, erect a perpendicular, and lay off a distance JK equal to the distance between centres, C, multiplied by the experimentally determined constant 0·314; about the point K so determined, draw a circular arc AB, tangent to the line GH. Any line drawn tangent to this arc will be the common tangent to a pair of cone steps giving the same belt length as that of the given pair. For example, suppose that OD is the radius of one step of the required pair; about O, with a radius equal to OD, draw a circle; tangent to this circle and to the arc AB, draw a straight line DE; about O' and tangent to DE, draw a circle; its diameter will equal that of the required step.

But suppose that instead of having one step of the required pair given, to find the other corresponding as above, a pair of steps are required that shall transmit a certain velocity ratio, $= r$, with the same length of belt as the given pair. Suppose OD and $O'E$ to represent the unknown steps. The given velocity ratio equals r. But from similar triangles $CD \div O'E = FO \div FO'$. Therefore $r = \dfrac{FO}{FO'}$; but $FO = C + x$, and $FO' = x$. Therefore $r = \dfrac{C + x}{x}$, and $x = \dfrac{C}{r - 1}$. Hence, with r and C given, the distance x may be found, such that if from F a line be drawn tangent to AB, the cone steps drawn tangent to it will give the velocity ratio, r, and a belt length equal to that of any pair of cones determined by a tangent to AB. The point F often falls at an inconvenient distance. The radii of the required steps may then be found as follows: Place a straight-edge tangent to the arc AB and measure the perpendicular distances from it to O and O'. The straight-edge may be shifted until these distances bear the required relation to each other.

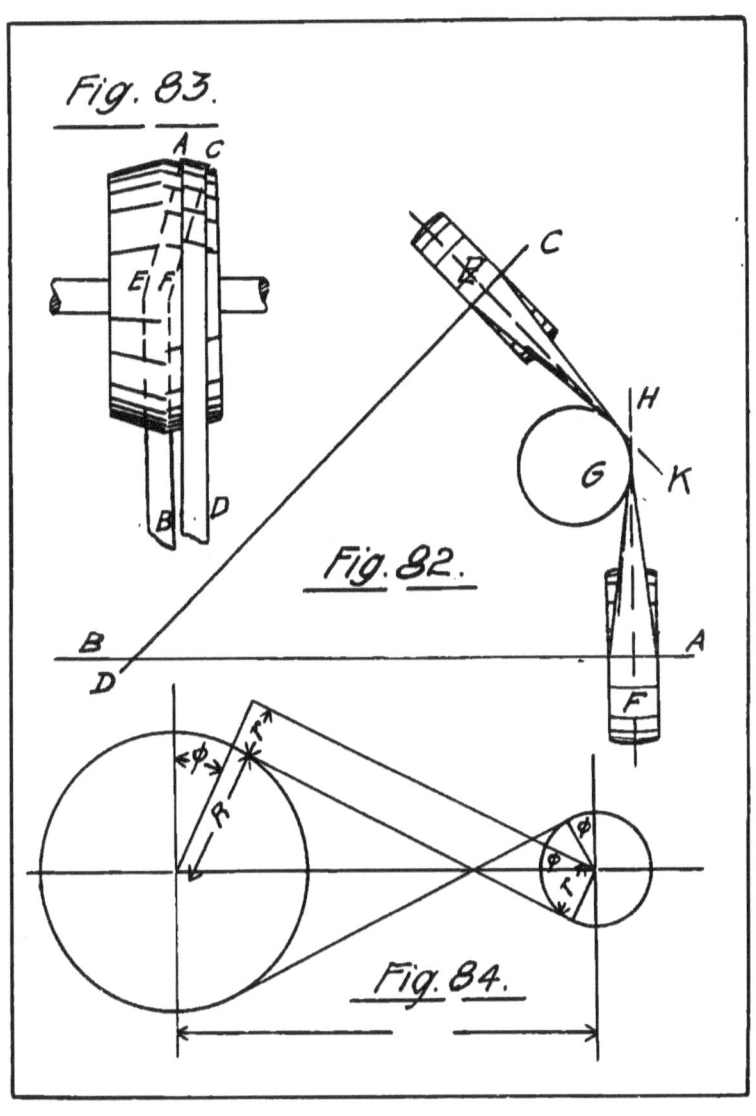

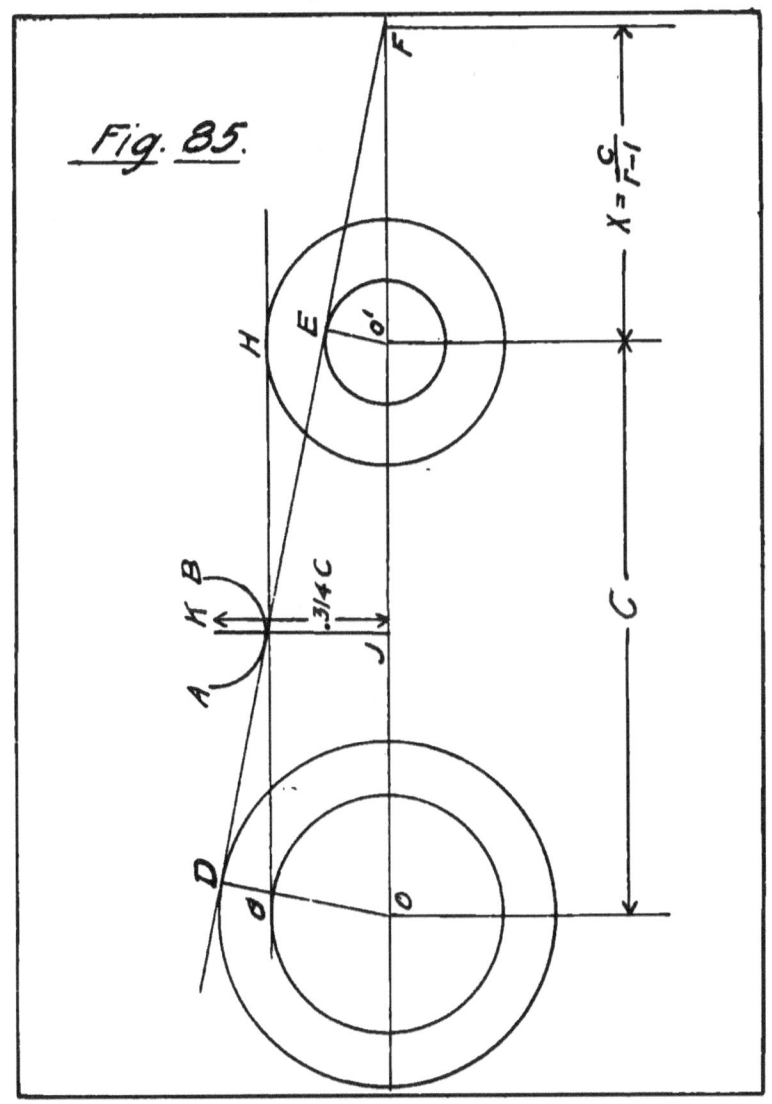

Fig. 85.

BELTS. 81

81. Design of Belts. — Fig. 86 represents two pulleys connected by a belt. When no moment is applied tending to produce rotation this tension in the two sides of the belt is equal. Let T_3 represent this tension. If now an increasing moment, represented by Rl, be applied to the driver, its effect is to increase the tension in the lower side of the belt and to decrease the tension in the upper side. With the increase of Rl this difference of tension increases till it is equal to P, the force with which rotation is resisted at the surface of the pulley. Then rotation begins,* and continues as long as this equality continues; *i. e.*, as long as $T_1 - T_2 = P$, in which $T_1 =$ tension in the driving side, and $T_2 =$ tension in the slack side. The tension in the driving side is increased at the expense of that in the slack side. Therefore $\dfrac{T_1 + T_2}{2} = T_3$.

To find the value of $\dfrac{T_1}{T_2}$. The increase in tension from the slack side to the driving side is possible because of the frictional resistance between the belt and pulley surface. Consider any element of the belt, ds, Fig. 88 (*a*). It is in equilibrium under the action of the following forces: T, the value of the varying tension corresponding to the section, acts upon one end of ds and is aided by dF. The force $T + dT$ acts upon the other end. From the action of these forces there results a normal pressure between ds and pulley, $= pds$, in which $p =$ the pressure per linear unit of belt. Draw the force triangle, (*b*) Fig. 88. It is an isosceles triangle, and hence $pds = (T + dT)\theta$; but $\theta = \dfrac{ds}{r}$; $\therefore pds = \dfrac{(T + dT)ds}{r}$; dT vanishes; $\therefore p = \dfrac{T}{r}$.

Since the force triangle is an isosceles triangle, it follows that $T + dT = T + dF$; hence $dT = dF$. Suppose that rotation occurs and that the belt slips upon the pulley at a rate corresponding to a

---

* While the moving parts are being brought up to speed the difference of tension must equal $P +$ force necessary to produce the acceleration.

82   MACHINE DESIGN.

coefficient of friction f. Then $dF = fpds$, and since $p = \dfrac{T}{r}$,

$$\therefore dT = f\dfrac{Tds}{r}; \text{ but } ds = rd\theta,$$

$$\therefore dT = fTd\theta;$$

$$\dfrac{dT}{T} = fd\theta;$$

$$\int_{T_2}^{T_1}\dfrac{dT}{T} = f\int_0^a d\theta;$$

$$\log_e \dfrac{T_1}{T_2} = fa;$$

$\dfrac{T_1}{T_2} = e^{fa}$, where $e =$ the Naperian base.

$$\log \dfrac{T_1}{T_2} = 0.4343 fa.$$

a is in π measure and equals a in degrees $\times$ 0·0174.

82. The following equations are established :

$$T_1 - T_2 = P \tag{1}$$

$$T_1 + T_2 = 2T_3. \tag{2}$$

$$\dfrac{T_1}{T_2} = e^{fa} \text{ or } \log \dfrac{T_1}{T_2} = 0.4343\,fa \tag{3}$$

The right hand members of (1) and (3) can usually be determined; hence the value of T_1 (the maximum stress in the belt) may be found, and proper proportions may be given to the belt. If W foot-pounds per minute are to be transmitted, and the velocity of the rim of the pulley transmitting this power in feet per minute equal S, then the force P equals the work divided by the velocity; or, $P = \dfrac{W}{S}$. The value of a is found from the diameters of the pulleys and their distance between centres, and may usually be estimated accurately enough. The value of f, the coefficient of friction, varies with the kind of belting, the material and character of surface of the pulley, and with the rate of slip of the belt on the pulley. Experiments made at the laboratory of the Massachusetts Institute of

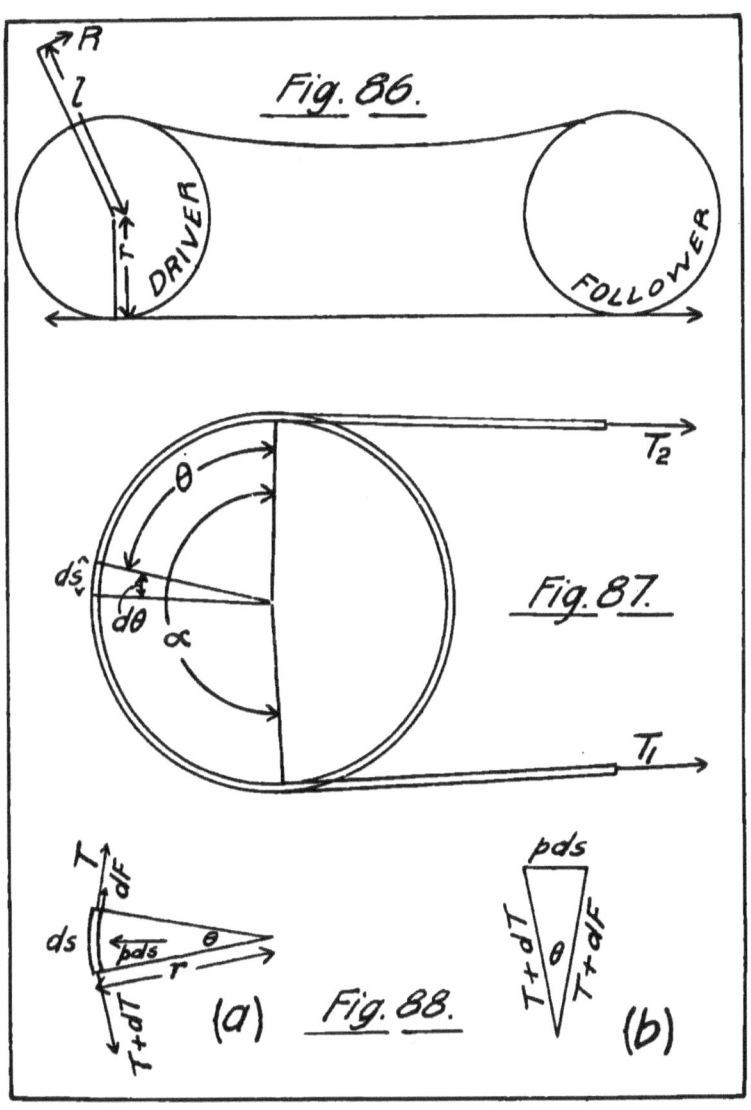

Technology, under the direction of Professor Lanza, indicate that for leather belting running on turned cast iron pulleys, the rate of slip for efficient driving is about three to four feet per minute ; and also that the coefficient of friction corresponding to this rate of slip is about 0·27. The value 0·3 may be used. If this value of f be used the slip will be kept within the above limits if the belt be put on with a proper initial tension, $= T_3 = \dfrac{T_1 + T_2}{2}$, and the driving of the belt so designed will be satisfactory.

83. *Problem*. — A single-acting pump has a plunger $8'' = 0\cdot666$ feet in diameter, whose stroke has a constant length of $10'' = 0\cdot833$ feet. The number of strokes per minute is 50. The plunger is actuated by a crank, and the crank shaft is connected by spur gears to a pulley shaft, the ratio of gears being such that the pulley shaft runs 300 revolutions per minute. The pulley which receives the power from the line shaft is 18" in diameter. The pressure in the delivery pipe is 100 lbs. per square inch. The line shaft runs 150 revolutions per minute, and its axis is at a distance of 12 feet from the axis of the pulley shaft. Since the line shaft runs half as fast as the pulley shaft, the diameter of the pulley on the line shaft must be twice as great as that on the pulley shaft, or 36". The work to be done per minute, neglecting the friction in the machine, is equal to the number of pounds of water pumped per minute multiplied by the head in feet against which it is pumped. The number of cubic feet of water per minute equals the displacement of the plunger in cubic feet multiplied by the number of strokes per minute $= \dfrac{0.666^2 \times \pi}{4} \times 0\cdot833 \times 50 = 14\cdot5$, and therefore the number of pounds of water pumped per minute $= 14\cdot5 \times 62\cdot4 = 907$. One foot vertical height or "head" of water corresponds to a pressure of $0\cdot435$ lbs. per square inch, and therefore 100 lbs. per square inch corresponds to a "head" of $100 \div 0\cdot435 = 230$ feet. The work done per minute in pumping the water therefore is equal to 907 lbs. $\times$ 230 feet $=$ 208,610 foot-pounds. The velocity of the rim of the belt pulley $= 300 \times 1\cdot5\pi = 1410$ feet per minute. Therefore the force $P = T_1 - T_2$. 208,610 ft.lbs. per minute ÷ 1410 feet per minute = 147 lbs.

To find u, see Fig. 89. $\sin \theta = \dfrac{R-r}{l} = \dfrac{9''}{144} = 0·0625$. Therefore $\theta = 3° 35'$; $u = 180° - 2\theta = 180° - 7° 10' = 173°$ nearly; u in π measure $= 173 \times 0·0174 = 3·01$.

$$\log \frac{T_1}{T_2} = 0·4343 \times f \times u = 0·4343 \times 0·3 \times 3·01 = 0·3921.$$

$$\therefore \frac{T_1}{T_2} = 2·46 \; ; \; P = T_1 - T_2 = 147.$$

Combining these equations T_1 is found to be equal to 246 lbs., = the maximum stress in the belt. Experiment shows that 70 lbs. per inch of width of a *laced, single belt* is a safe working stress. Therefore the width of the belt = 246 ÷ 70 = 3·5". The friction of the machine might have been estimated and added to the work to be done.

84. *Problem.* — A sixty horse-power dynamo is to run 1500 revolutions per minute, and has a 15" pulley on its shaft. Power is supplied by a line shaft running 150 revolutions per minute. A suitable belt connection is to be designed. The ratio of angular velocities of dynamo shaft to line shaft is 10 to 1; hence the diameter of the pulley on the line shaft would have to be ten times as great as that of the one on the dynamo, = 12·5 feet, if the connections were direct. This is inadmissible, and therefore the increase in speed must be obtained by means of an intermediate, or *counter shaft*. Suppose that the diameter of the largest pulley that can be used on the counter shaft = 48". Then the necessary speed of the counter shaft $= 1500 \times \dfrac{15}{48} = 470$ nearly. The ratio of diameters of the required pulleys for connecting the line shaft and the counter shaft $= \dfrac{470}{150} = 3·13$. Suppose that a 60" pulley can be used on the line shaft; then the diameter of the required pulley for the counter shaft will $= \dfrac{60}{3·13} = 19"$ nearly. Consider first the belt to connect

TABLE II.—For Use in Designing Belts.

Width of Belt	120°	125°	130°	135°	140°	145°	150°	155°	160°	165°	170°	175°	180°	185°	190°	195°	200°
1·5 in.	49	50	52	53	54	56	57	58	59	61	62	63	64	65	66	67	68
2·0	65	67	69	71	72	74	76	78	79	81	83	84	85	87	88	90	91
2·5	81	83	86	88	90	92	95	97	99	101	103	105	106	108	110	112	113
3·0	97	100	103	106	109	111	114	116	118	121	124	126	128	130	132	134	136
3·5	114	117	120	123	126	129	133	135	138	141	144	147	149	151	154	156	158
4·0	130	134	137	141	144	148	152	155	158	161	164	168	171	173	176	179	181
4·5	146	150	154	158	162	166	170	174	178	181	185	188	192	194	198	201	204
5·0	162	167	172	176	181	185	190	193	197	201	205	209	213	216	220	223	227
5·5	178	184	189	194	199	203	208	212	217	221	226	230	234	238	242	246	249
6·0	195	201	207	212	217	222	227	232	237	242	246	251	255	260	264	268	272
6·5	211	217	224	230	235	240	246	251	257	262	267	272	276	281	286	290	294
7·0	227	234	241	247	253	259	265	271	276	282	287	293	298	303	308	313	317
7·5	243	251	258	264	271	277	284	290	296	303	308	314	320	325	330	335	339
8·0	260	268	276	282	290	296	303	310	316	323	328	335	342	346	352	358	362
8·5	276	284	293	300	308	315	322	329	336	343	349	356	363	368	374	380	385
9·0	292	301	310	318	326	333	340	348	356	363	370	377	384	390	396	402	408
9·5	308	317	327	335	344	352	359	367	376	383	390	398	405	411	418	424	431
10·0	325	334	344	353	362	371	378	387	396	404	411	419	426	433	440	447	454
10·5	341	351	361	370	380	389	396	406	415	424	432	440	447	455	462	469	477
11·0	358	368	378	387	398	407	416	426	435	444	453	461	469	477	484	492	500
11·5	374	385	395	405	416	425	435	445	454	464	473	481	490	498	506	515	522
12·0	390	402	413	423	435	445	454	465	474	484	494	502	511	520	528	537	545
12·5	406	418	430	441	453	463	473	484	494	505	514	523	532	541	550	559	567
13·0	422	435	447	459	471	482	492	504	515	525	535	544	553	563	572	581	590
13·5	439	452	465	477	489	501	511	524	535	546	555	565	575	585	595	604	613
14·0	455	469	482	495	507	519	530	542	553	565	577	586	595	606	616	626	636
14·5	471	486	499	512	525	538	550	562	575	586	598	607	618	628	638	649	659
15·0	488	503	517	530	544	557	569	582	595	607	619	629	640	650	660	672	682
15·5	504	520	534	548	562	575	588	601	615	627	639	650	662	672	683	694	704
16·0	521	537	551	566	580	594	607	621	634	647	659	671	683	694	705	716	726
16·5	537	553	568	583	598	613	626	641	654	667	680	692	704	715	727	739	749
17·0	552	570	586	601	617	632	645	660	674	687	700	713	725	737	749	762	772
17·5	567	586	603	619	634	650	664	679	693	707	720	734	746	759	771	784	795
18·0	583	602	621	637	652	668	683	699	713	727	741	755	767	781	793	806	817
18·5	597	619	638	654	670	686	702	718	733	748	761	776	789	803	815	828	839
19·0	619	637	655	672	689	705	721	737	753	768	782	796	811	825	837	851	862
19·5	635	653	672	689	707	723	740	756	773	788	802	817	832	846	859	873	885
20·0	651	670	689	707	725	742	759	776	792	808	823	838	853	867	881	895	908

Angle of Contact a.

the dynamo to the counter shaft. The work $= 60 \times 33,000 = 1,980,000$ foot-pounds per minute; the rim of the dynamo pulley moves $\frac{\pi 15}{12} \times 1500 = 5890$ feet per minute. Therefore $T_1 - T_2 = \frac{1,980,000}{5890} = 336$ lbs. The axis of the counter shaft is 10 feet from the axis of the dynamo, and as before $\sin \theta = \frac{R-r}{l} = \frac{24 - 7.5}{120} =$ 0·1378. Therefore $\theta = 8°$ nearly.

$$a = 180° - 2\theta = 164°$$

a in π measure $= 164 \times 0·0174 = 2·85.$

$$\log \frac{T_1}{T_2} = 0·4343 \times 0·3 \times 2·85 = 0·3713,$$

$$\frac{T_1}{T_2} = 2·35.$$

From these equations $T_1 = 583$ lbs., and the safe width of the single belting $= 583 \div 70 = 8·34''$; say 8·5''. The width of the belt to connect the line shaft to the counter shaft may be found by the same method.

85. Table II is given to save the above calculations for each belt. The body of the table is made up of values of P, the driving force at the pulley surface. To use the table, suppose that the smaller angle of contact of the belt with the two pulleys considered is known, $= a°$. P is also known. Find $a°$, or the nearest smaller value, in the horizontal column at the head of the table. In the vertical column under this value of a, find P, or the next greater value. Horizontally opposite this in the first vertical column, is the safe width of a single belt. If a double belt is to be used the value found may be divided by 2.

86. From equation (3), p. 82, it follows that the *ratio* of tensions, $\frac{T_1}{T_2}$, when the belt slips at a certain allowable rate (*i. e.*, when f is constant) depends only upon a. This ratio, therefore, is independent of the initial tension, T_2; hence "taking up" a belt does not change $\frac{T_1}{T_2}$. The *difference* of tension, $T_1 - T_2 = P$, is, however, dependent on T_2. Because p, the normal pressure between belt and pulley, varies directly as T_2. Then since $dF = fp\,ds = dT$, it follows that dT varies with T_2, and hence

$$\int dT = T_1 - T_2 = P$$

varies with T_2. This is equivalent to saying that "taking up" a belt increases its driving capacity, and "letting it out" decreases its driving capacity.

This result is modified because another variable enters the problem. If T_2 be changed, the amount of slipping changes, and the coefficient of friction varies directly with the amount of slipping. Therefore, an increase of T_2 would increase p and decrease f in the expression $fp\,ds = dT$, and the converse is also true. This is probably of no practical importance.

The value of P may also be increased by increasing either f, the coefficient of friction, or a, the arc of contact; since increase of either increases the ratio $\frac{T_1}{T_2}$; and therefore increases $T_1 - T_2 = P$.

Increasing T_2 decreases the life of the belt. It also increases the pressure on the bearings in which the pulley shaft runs, and therefore increases frictional resistance; hence a greater amount of the energy supplied is converted into heat and lost to any useful purpose. But if T_2 be kept constant and f or a be increased, the driving power is increased without increase of pressure in the bearings, because this pressure $= 2T_2 =$ constant. When possible, therefore, it is preferable to increase P by increase of f or a, rather than by increase of T.

Application of belt dressing may serve sometimes to increase f.

87. If, as in Fig. 86, the arrangement is such that the upper side of the belt is the slack side, the "sag" of the belt tends to increase the arc of contact, and therefore to increase $\frac{T_1}{T_2}$. If the lower side is the slack side, the belt sags away from the pulleys and a and $\frac{T_1}{T_2}$ are decreased.

An idler pulley, C, may be used, as in Fig. 90. It is pressed against the belt by some means. Its purpose may be to increase P by increasing the tension, $T_3, = \frac{T_1 + T_2}{2}$. In this case friction in the bearings is increased. Or it may be used on a slack belt to increase the angle of contact, a, the ratio $\frac{T_1}{T_2}$, and therefore P, the driving force. In this case the value of $T_3, = \frac{T_1 + T_2}{2}$, may be made as small a value as is consistent with driving, and hence the journal friction may be small.

Tighteners are sometimes used with slack belts for disengaging gear, the driving pulley being vertically below the follower.

In the use of any device to increase f and a, it should be remembered that T_1 is thereby increased, and may become greater than the value for which the belt was designed. This may result in injury to the belt.

In Fig. 91, the smaller pulley, A, is above the larger one, B. A has a smaller arc of contact, and hence the belt would slip upon it sooner than on B. The weight of the belt, however, tends to increase the pressure between the belt and A, and to decrease the pressure between the belt and B. The driving capacity of A is thereby increased, while that of B is diminished; or, in other words, the weight of the belt tends to equalize the inequality of driving power. If the larger pulley had been above, there would have been a tendency for the belt weight to increase the inequality of driving capacity of the pulleys. The conclusion from this, as to arrangement of pulleys, is obvious.

88. A belt resists a force which tends to bend it. Work must be done, therefore, in bending a belt around a pulley. The more it is bent the more work is required. Suppose AB, Fig. 92, to represent a belt which moves from A toward B. If it runs upon C it must

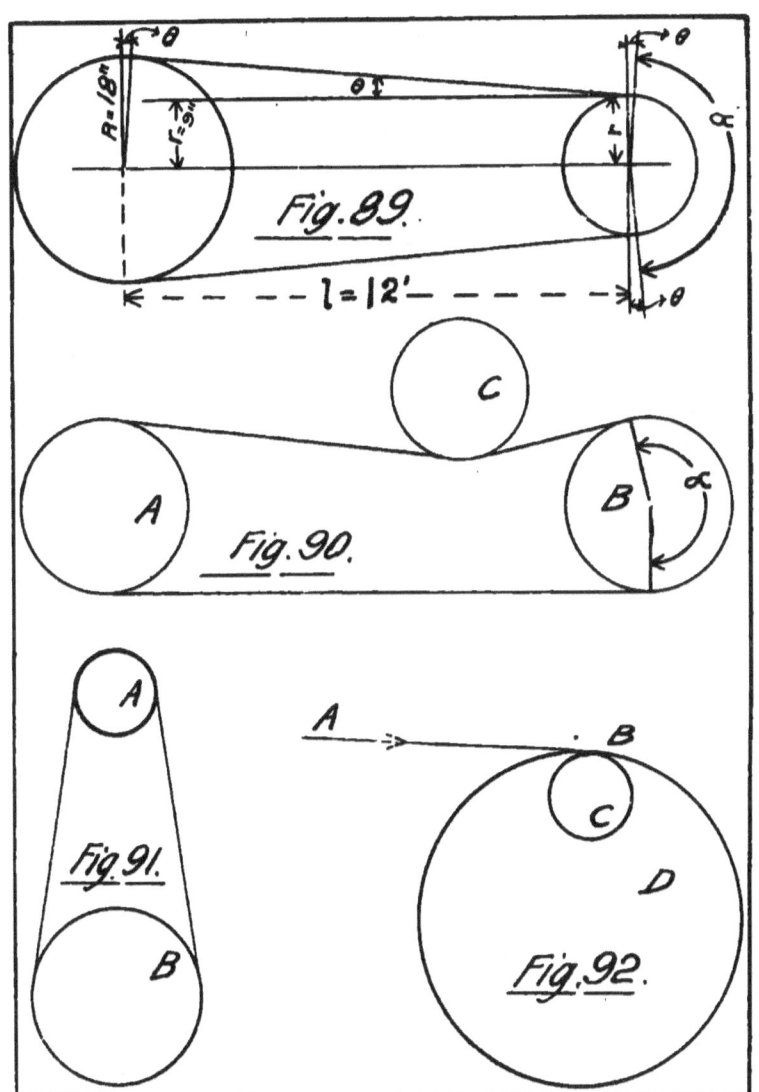

be bent more than if it runs upon D. The work done in bending the belt is converted into useless heat by the friction between the belt fibres. It is desirable, therefore, to do as little bending as possible. This is one reason why large pulleys in general are more efficient than small ones. The resistance to bending increases with the thickness of the belt, and hence double belts should not be used on small pulleys if it can be avoided.

89. Effect of Centrifugal Force of Belts. — In Fig. 93, as the belt reaches a, it has its direction of motion changed. The belt tends to move on in a straight line, and therefore resists the change of direction. There results a force acting radially outward, which *tends* to cause the belt to leave the pulley. The measure of this force per linear inch of belt $= c = \dfrac{w}{g} \times \dfrac{v^2}{r}$; in which $w =$ weight of belting per linear inch, $v =$ belt velocity in feet per second, $g = 32 \cdot 2$ feet per second acceleration, $r =$ pulley radius. As the velocity of the belt is increased, w and r remaining constant, c will increase, and will eventually equal the radial pressure at a per linear inch of the belt $= p$. With further increase of v the belt would leave the pulley at a. This would result in a decrease in the arc of contact, and hence a decrease in the driving capacity of the belt. This centrifugal force is the same for every linear inch of the belt which is in contact with the pulley, *i. e.*, the radial force acting outward is constant from a around to b. The radial pressure p, acting inward, however, increases from a around to b; hence the tendency to leave the pulley is greatest at a. Experience shows that this may occur in practice, as shown in Fig. 94, the angle of contact being reduced from a to β. The value of the radial pressure, $= p$, due to belt tension, at a, the middle of the first inch of contact, may be found. The value is less than for any other inch of contact, because it increases from a around to b. This value compared with the centrifugal force found as above shows the tendency for the belt to leave the pulley.

To find p. — In Fig. 95 the first inch of belt in contact with the pulley at a, Fig. 93, is represented. This element of the belt subtends an angle θ, whose value depends on the radius of the pulley.

90 MACHINE DESIGN.

The element of the belt is in equilibrium under the influence of three forces, $T_2 + \Delta T_2$, $T_2 + \Delta F$, and the reaction of p; i. e., the pressure of the pulley against the element of the belt. This reaction = p, and its line of action (being radial) makes equal angles with the lines of action of $T_2 + \Delta T_2$ and $T_2 + \Delta F$. The angle between the lines of action of these equal forces ($T_2 + \Delta T_2$ and $T_2 + \Delta F$) = θ. The force triangle is therefore the isosceles triangle shown. In this triangle

$$\frac{p}{T_2 + \Delta F} = \frac{\sin\theta}{\sin\beta}; \text{ but } \beta = \frac{180° - \theta}{2}, \text{ and } \Delta F = \frac{T_1 + T_2}{s},$$

in which s = number of inches of contact of belt with pulley. Therefore

$$p = \frac{T_2 + \dfrac{T_1 - T_2}{s}\sin\theta}{\sin\dfrac{180° - \theta}{2}}.$$

Apply this to the problem on page 84: $T_1 = 583$; $T_2 = \dfrac{T_1}{2\cdot 36} = 247$.

$$T_1 - T_2 = 336.$$

$$s = a \times r = 164° \times 0\cdot 0174 \times 7\cdot 5 = 21\cdot 4''.$$

θ = angle subtended by 1″ on 7·5″ radius = $360° \times \dfrac{1}{2 \times 7\cdot 5\pi} = 7\cdot 64°$
= 7° 38′. Sin $\theta = 0\cdot 1328$.

$$\frac{180° - \theta}{2} = \frac{172° \; 22'}{2} = 86° \; 11';$$

$$\sin\frac{180° - \theta}{2} = 0\cdot 9977;$$

$$p = \frac{\left(247 + \dfrac{366}{21\cdot 4}\right) 0\cdot 1328}{0\cdot 9977} = 262\cdot 7 \times \frac{0\cdot 01328}{0\cdot 9977} = 34\cdot 7.$$

To find c. — A cubic inch of belting weighs about 0·04 lb. Single belting is about 0·25″ thick, and in this case the width is 8·5″. The weight of belt per linear inch is therefore $W = 0\cdot 25 \times 8\cdot 5 \times 0\cdot 04 = 0\cdot 085$ lbs.

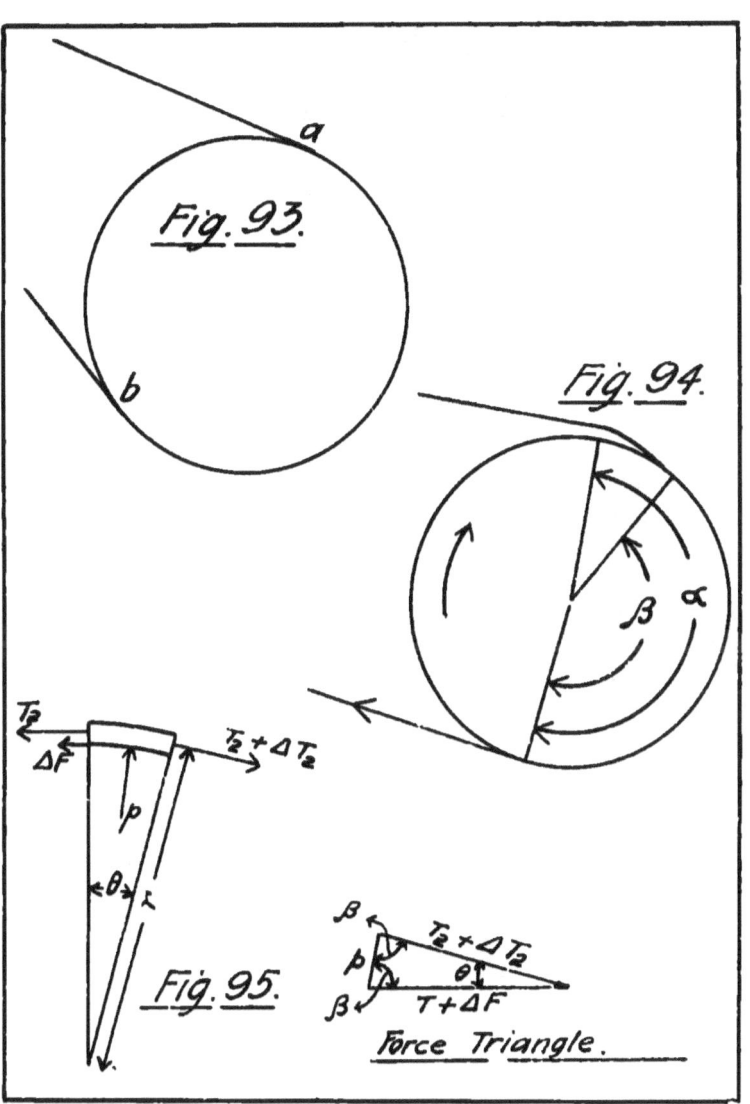

BELTS. 91

$$v = \frac{15\pi \times 1500}{12 \times 60} = 98\cdot 2 \text{ feet per second.}$$

$$c = \frac{wv^2}{gr} = \frac{0\cdot 085 \times 98\cdot 2^2}{32\cdot 2 \times \frac{7\cdot 5}{12}} = 40\cdot 8 \text{ lbs.}$$

The centrifugal force is in excess at (a) the middle of the first inch of contact, therefore, by an amount equal to 40·8 lbs. — 34·7 lbs. = 6·1 lbs. There would be a tendency for the belt to lift. This is opposed, however, by the weight of the belt. If the slack side of the belt be supposed to be straight and horizontal, one-half its weight will be supported at a. The distance between centres of shafts = 10 ft. = 120". The weight W of the slack side = cubic contents × weight per cubic unit, = 0·25 × 8·5 × 120 × 0·04 = 10·2 lbs. = W. Half of this aids p in its opposition to c at a. Hence the outward radial force at a in this case = 40·8 — (34·7 + 5·1) = 1 lb. If the direction of rotation were reversed the slack side would be below and the outward radial force at b would equal 40·8 — (34·7 — 5·1) = 11·2 lbs. If the line joining the centres is inclined, as in Fig. 97, only the component of W at right angles to the belt, = P, is effective to produce inward radial pressure at a. If the slack side of the belt becomes vertical P becomes = 0, and hence the weight has no effect.

To find the velocity of belt at which, under given conditions, the belt just tends to leave the pulley. — Let w = radial force at a due to belt weight. The belt just tends to leave the pulley when the sum of the inward radial forces = the sum of the outward radial forces; or when $c = p \pm w$; substituting value of $c = \frac{Wv^2}{gr}$, in which W = weight of one lineal inch of the belt, and solving for $v =$ $\sqrt{\frac{(p \pm w) gr}{W}}$. If a belt tends to leave the pulley, running under given conditions, it would seem that increasing the radius of the pulley upon which the slack side runs would reduce the centrifugal force, since $c \propto \frac{1}{r}$; but in order to keep the shaft running at the same rate as before, the belt must run faster and $c \propto v^2$. Also, since the moment to produce rotation is constant, the force (with the increase of lever arm due to increased size of pulley) is less, and hence (unless a wider belt than is necessary is used) the width of the belt and hence its weight per linear inch must be reduced, and $c \propto w$.

In the design of belting care should be taken not to make the distance between the shafts carrying the pulleys too small, especially if there is the possibility of sudden changes of load. Belts have some elasticity, and the total yielding under any given stress is proportional to the length, the area of cross-section being the same. Therefore a long belt becomes a yielding part, or spring, and its yielding may reduce the stress due to a suddenly applied load to a safe value; whereas in the case of a short belt, with other conditions exactly the same, the stress due to much less yielding might be sufficient to rupture or weaken the joint.

CHAPTER VIII.

DESIGN OF FLY-WHEELS.

90. Often in machines there is capacity for uniform effort, but the resistance fluctuates. In other cases a fluctuating effort is applied to overcome a uniform resistance, and yet in both cases a more or less uniform rate of motion must be maintained. When this occurs, as has been explained, a moving body of considerable weight is interposed between effort and resistance, which, because of its weight, absorbs and stores up energy with increase of velocity when the effort is in excess, and gives it out with decrease of velocity when the resistance is in excess. This moving body is usually a rotating body, called a fly-wheel.

To fulfill its office a fly-wheel must have a variation of velocity; because it is by reason of this variation that it is able to store and give out energy. The kinetic energy, E, of a body whose weight is W, moving with a velocity v, is expressed by the equation

$$E = \frac{Wv^2}{2g}.$$

To change E, with W constant, v must vary. The allowable variation of velocity depends upon the work to be accomplished. Thus, the variation in an engine running electric lights, or spinning machinery, should be very small; probably not greater than a half of one per cent. While a pump or a punching machine may have a much greater variation without interfering with the desired result. If the maximum velocity, v_1, of the fly-wheel rim, and the allowable variation are known, the minimum velocity, v_2, becomes known; and the energy that can be stored and given out with the allowable change of velocity is equal to the difference of kinetic energy at the two velocities.

$$E = \frac{Wv_1^2}{2g} - \frac{Wv_2^2}{2g} = \frac{W}{2g}(v_1^2 - v_2^2).$$

The general method for fly-wheel design is as follows: Find the maximum energy due to excess or deficiency of effort during a cycle of action, $= E$. Assume a convenient mean diameter of fly-wheel rim. From this and the given maximum rotative speed of the fly-wheel shaft, find v_1. From v_1 and the given allowable variation of velocity, find v_2. Solve the above equation for W, thus:

$$W = \frac{2gE}{v_1^2 - v_2^2}.$$

Substitute the values of $E, v_1, v_2,$ and $g = 32.2$ ft. per second. Whence W becomes known, $=$ weight of fly-wheel rim. The weight of rim only will be considered; the other parts of the wheel being nearer the axis have less velocity, and less capacity per pound for storing energy. Their effect is to reduce slightly the allowable variation of velocity.

91. *Problem*. — In a punching machine the belt is *capable* of applying a uniform torsional effort to the shaft; but most of the time it is only required to drive the moving parts of the machine against frictional resistance. At intervals, however, the punch must be forced through metal which offers shearing resistance to its action. Either the belt or fly-wheel, or the two combined, must be capable of overcoming this resistance. A punch makes 30 strokes per minute, and enters the die $\frac{1}{4}"$. It is required to punch $\frac{3}{4}"$ holes in steel plates $\frac{1}{2}"$ thick. The shearing strength of the steel is about 50,000 pounds per square inch. When the punch just touches the plate the surface which offers shearing resistance to its action equals the surface of the hole which results from the punching, $= \pi d t$, in which $d =$ diameter of hole or punch, $t =$ thickness of plate. The maximum shearing resistance, therefore, equals $\pi \frac{3}{4} \times \frac{1}{2} \times 50000 = 58800$ lbs. As the punch advances through the plate the resistance decreases, because the surface in shear decreases, and when the punch just passes through the resistance becomes zero. If the change of resistance be assumed uniform (which would probably be approximately true) the mean resistance to punching would equal the maximum resistance + minimum resistance, $\div 2, = \dfrac{58800 + 0}{2}$ = 29400. The radius of the crank which actuates the punch = 2". In Fig. 98 the circle represents the path of the crank-pin centre. Its vertical diameter then represents the travel of the punch. If the actuating mechanism be a slotted cross-head, as is usual, it is a case of harmonic motion, and it may be assumed that while the punch travels vertically from A to B, the crank-pin centre travels in the semicircle ACB. Let BD and DE each $= \frac{1}{4}$ inch. Then when the punch reaches E it just touches the plate to be punched, which is $\frac{1}{4}$" thick, and when it reaches D it has just passed through the plate. Draw the horizontal lines EF and DG and the radial lines OG and OF. Then, while the punch passes through the plate, the crank-pin centre moves from F to G, or through an angle (in this case) of 19°. Therefore the crank shaft A, Fig. 99, and attached gear rotate through 19° during the action of the punch. The ratio of angular velocity of the pinion and the gear = the inverse ratio of pitch diameters $= \dfrac{60}{12} = 5$. Hence the shaft B rotates through an angle $= 19° \times 5 = 95°$ during the action of the punch. If there were no fly-wheel the belt would need to be designed to overcome the maximum resistance; i. e., the resistance at the instant when the punch is just beginning to act. This would give for this case a double belt about 20" wide. The need for a fly-wheel is therefore apparent. Assume that the fly-wheel may be conveniently 36" mean diameter, and that a single belt 5" wide is to be used. The allowable maximum tension is then $= 5 \times$ allowable tension per inch of width of single belting $= 5 \times 70 = 350$ lbs. $= T_1$. Then from the equation $\dfrac{T_1}{T_2} = e^{fa}$, if $a° = 180°$, $\dfrac{T_1}{T_2} = 2\cdot56$; hence $T_2 = \dfrac{T_1}{2\cdot56} = \dfrac{350}{2\cdot56} =$ 136·5, and $T_1 - T_2 = 213\cdot5$ lbs. = the driving force at the surface of the pulley. Assume that the frictional resistance of the machine is equivalent to 25 lbs. applied at the pulley rim. Then the belt can exert 213·5 lbs. $- 25 = 188\cdot5$ lbs., $= P$, to accelerate the fly-wheel or to do the work of punching. Assume variation of velocity $= 10$ per cent. The work of punching $=$ the mean resistance offered to the punch multiplied by the space through which the punch acts, $= \dfrac{58800}{2} \times 0\cdot5 = 14700$ inch-pounds $= 1220$ foot-pounds. The pulley shaft moves during the punching through 95°, and the driving tension of the belt, $= P = 188\cdot5$ lbs., does work $= P \times$ space moved through during the punching $= 188\cdot5$ lbs. $\times \pi d \dfrac{95°}{360} = 188\cdot5$ lbs. $\times \pi \times 2$ ft. $\times \dfrac{95}{360} = 311$ foot-pounds. The work left for the fly-wheel to give out with a reduction of velocity of 10 per cent. $= 1220 - 311 = 909$ foot-pounds. Let $v_1 =$ maximum velocity of fly-wheel rim; $v_2 =$ minimum velocity of fly-wheel rim; $W =$ weight of the fly-wheel rim. The energy it is capable of giving out, while its velocity is reduced from v_1 to v_2, $= \dfrac{W(v_1^2 - v_2^2)}{2g}$, and the value of W must be such that this energy given out shall equal 909 foot-pounds. Hence the following equation may be written:

$$\dfrac{W(v_1^2 - v_2^2)}{2g} = 909.$$

Therefore $\qquad W = \dfrac{909 \times 2 \times g}{v_1^2 - v_2^2}.$

The punch shaft makes 30 revolutions per minute and the pulley shaft $30 \times 5 = 150 = N$ revolutions per minute. Hence v_1 in feet per second $= \dfrac{Nd\pi}{60}$; d being fly-wheel diameter in feet,

$$v_1 = \dfrac{150 \times 3\pi}{60} = 23\cdot5.$$

$$v_2 = 0\cdot90\, v_1 = 21\cdot1.$$

$$v_1^2 = 552; \quad v_2^2 = 446; \quad v_1^2 - v_2^2 = 106.$$

Hence $\qquad W = \dfrac{909 \times 2 \times 32\cdot2}{106} = 551$ lbs.

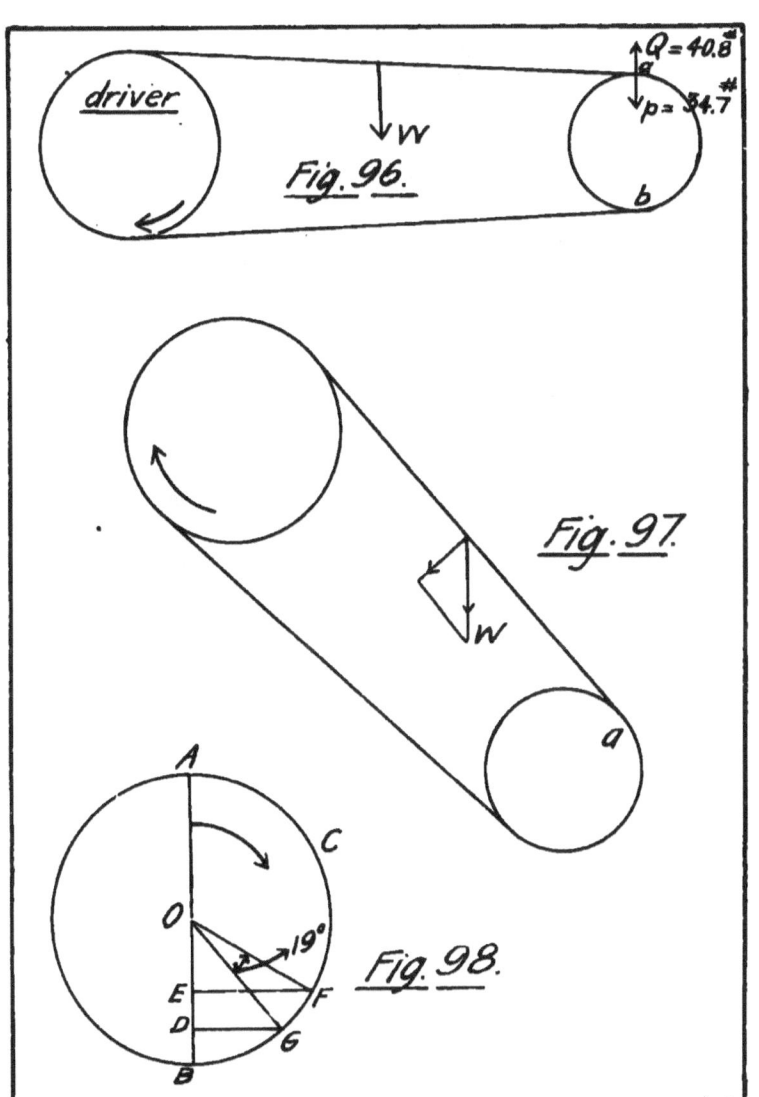

DESIGN OF FLY-WHEELS. 97

To proportion the rim.—A cubic inch of cast iron weighs 0·26 lbs.; hence there must be $\frac{551}{0·26} = 2120$ cu. in. The cubic contents of the rim = mean diameter $\times \pi \times$ its cross-sectional area, A, = 2120 cu. in.; hence $A = \frac{2120}{36'' \times \pi} = 18·8''$ sq. in.

If the cross-section were made square its side would = $\sqrt{18·8} = 4·34''$.

92. Pump Fly-Wheel.—The belt for the pump, p. 83, is designed for the *average* work. A fly-wheel is necessary to adapt the varying resistance to the capacity of the belt. The rate of doing work on the return stroke (supposing no resistance due to suction) is only equal to the frictional resistance of the machine. During the working stroke the rate of doing work varies because the velocity of the plunger varies, although the pressure is constant. The rate of doing work is a maximum when the velocity of the plunger is greatest. In Fig. 100, A is the velocity diagram; B is the force diagram; C is the tangential diagram drawn as indicated on pages 35-36. The belt, 3·5'' wide, is capable of applying a tangential force of 147 lbs. to the 18'' pulley rim. The velocity of the pulley rim = π 1·5 $\times$ 300 = 1410. The velocity of the crank-pin axis = π 0·833 $\times$ 50 = 130·8. Therefore the force of 147 lbs. at the pulley rim corresponds to a force = $147 \times \frac{1410}{130·8} = 1585$ lbs. applied tangentially at the crank-pin axis. This may be plotted as an ordinate upon the tangential diagram C, from the base line XX_1, using the same force scale. Through the upper extremity of this ordinate draw the horizontal line DE. The area between DE and XX_1 represents the work the belt is capable of doing during the working stroke. During the return stroke it is capable of doing the same amount of work. But this work must now be absorbed in accelerating the fly-wheel. Suppose the plunger to be moving in the direction shown by the arrow. From E to F the effort is in excess and the fly-wheel is storing energy. From F to G the resistance is in excess and the fly-wheel is giving out energy. The work the fly-wheel must be capable of giving out with the allowable reduction of velocity is that represented by the area under the curve above the line FG. From G to D, and during the entire return stroke, the belt is doing work to accelerate the fly-wheel. This work becomes stored kinetic energy in the fly-wheel. Obviously the following equation of areas may be written:

$$X_1EF + XGD + XHKX_1 = GMF.$$

The left hand member of this equation represents the work done by the belt in accelerating the fly-wheel; the right hand member represents the work given out by the fly-wheel to help the belt.

The work in foot-pounds represented by the area GMF may be equated with the difference of kinetic energy of the fly-wheel at maximum and minimum velocities. To find the value of this work: One inch of ordinate on the force diagram represents 4260 lbs.; one inch of abscissa represents 0·2245 ft. Therefore one square inch of area represents 4260 lbs. $\times$ 0·2245 = 956·37 foot-pounds. The area $GMF = 1·6$ sq. in. Therefore the work = 956·37 $\times$ 1·6 = 1530 foot-pounds $= E$. The difference of kinetic energy $= \dfrac{W}{g}(v_1^2 - v_2^2) = 1530$; W equals the weight of the fly-wheel rim. Hence

$$W = \frac{1530 \times 32·2}{v_1^2 - v_2^2}.$$

Assume the mean fly-wheel diameter = 2·5 ft. It will be keyed to the pulley shaft, and will run 300 revolutions per minute, = 5 revolutions per second. The maximum velocity of fly-wheel rim = 2·5$\pi \times 5 = 39·15 = v_1$. Assume an allowable variation of velocity, = 5 per cent. Then $v_2 = 39·15 \times 0·95 = 37·19$; $v_1^2 = 1532·7$; $v_2^2 = 1383·1$; $v_1^2 - v_2^2 = 149·6$. Hence

$$W = \frac{1530 \times 32·2}{149·6} = 329 \text{ lbs.}$$

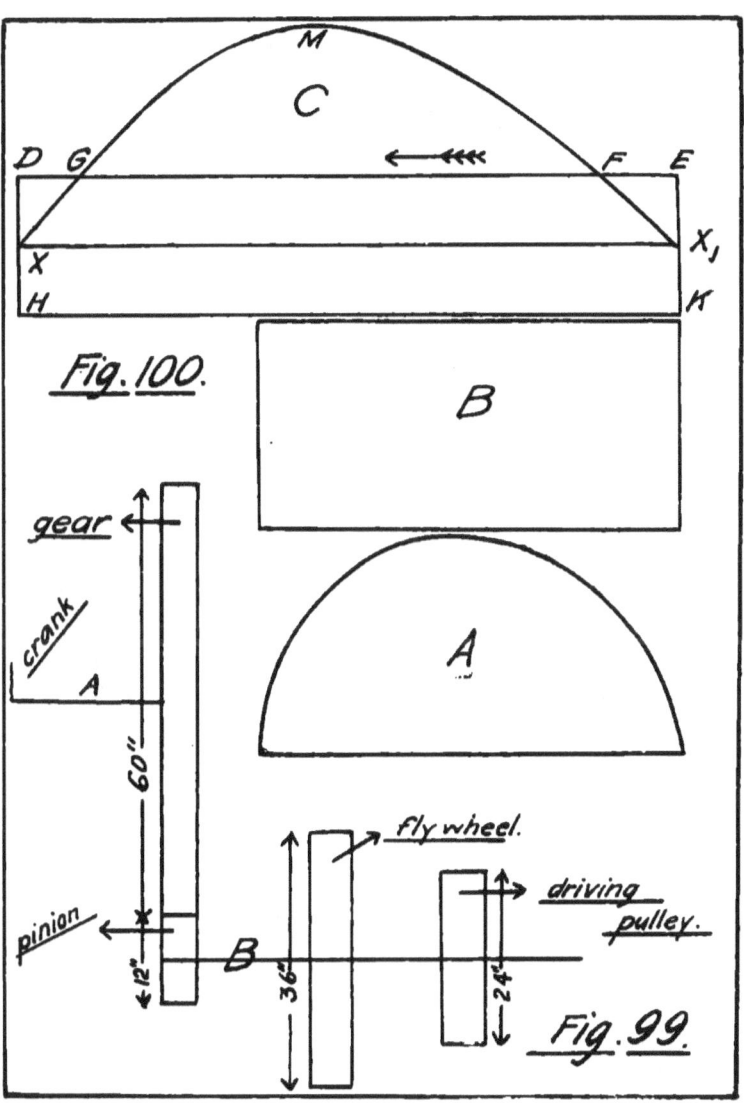

DESIGN OF FLY-WHEELS. 99

There must be 329 ÷ 0·26 cubic inches in the rim 1262. The pitch circumference = 30 × π = 94·2". Hence the cross-sectional area of rim = 1262 ÷ 94·2 = 13·4 +. The rim may be made 3" × 4·5".

The frictional resistance of the machine is neglected. It might have been estimated and introduced into the problem as a constant resistance.

93. Steam Engine Fly-Wheel. — From given data draw the indicator card as modified by the acceleration of reciprocating parts. See page 35 and Fig. 30. From this, and the velocity diagram, construct the diagram of tangential driving force, Fig. 31. Measure the area of this diagram and draw the equivalent rectangle on the same base. This rectangle represents the energy of the uniform resistance during one stroke; while the tangential diagram represents the work done by the steam upon the crank-pin. The area of the tangential diagram which extends above the rectangle represents the work to be absorbed by the fly-wheel with the allowable variation of velocity. Find the value of this in foot-pounds, and equate it to the expression for difference of kinetic energy at maximum and minimum velocity. Solve for W, the weight of fly-wheel.

CHAPTER IX.

RIVETED JOINTS.

94. A rivet is a fastening used to unite metal plates or rolled structural forms, as in boilers, tanks, built-up machine frames, etc. It consists of a head, A, Fig. 101, and a straight shank, B. It is inserted, usually red-hot, into holes, either drilled or punched in the parts to be connected, and the projecting end of the shank is then formed into a head (see dotted lines) either by hand or machine riveting. A rivet is a permanent fastening and can only be removed by cutting off the head. A row of rivets joining two members is called a *riveted joint* or *seam of rivets*. In hand riveting the projecting end of the shank is struck a quick succession of blows with hand hammers and formed into a head by the workman. A helper holds a sledge or "dolly bar" against the head of the rivet. In "button set" or "snap" riveting, the rivet is struck a few heavy blows with a sledge to "upset" it. Then a die or "button set," Fig. 102, is held with the spherical depression, B, upon the rivet; the head A is struck with the sledge, and the rivet head is thus formed. In machine riveting a die similar to B is held firmly in the machine and a similar die opposite to it is attached to the piston of a steam, hydraulic, or pneumatic cylinder. A rivet, properly placed in holes in the members to be connected, is put between the dies and pressure is applied to the piston. The movable die is forced forward and a head formed on the rivet.

The relative merits of machine and hand riveting have been much discussed. Either method carefully carried out will produce a good serviceable joint. If in hand riveting the first few blows be light the rivet will not be properly upset, the shank will be loose in the hole, and a leaky rivet results. If in machine riveting the axis of the rivet does not coincide with the axis of the dies, an off-set head results. See Fig. 103. In large shops where work must be turned out economically in large quantities, machines must be used. But there are always places inaccessible to machines, where the rivets must be driven by hand. Holes for the reception of rivets are usually punched, although for thick plates and very careful work they may be sometimes drilled. If a row of holes be *punched* in a plate, and a similar row as to size and spacing be *drilled* in the same plate, testing to rupture will show that the punched plate is weaker than the drilled one. If the punched plate had been annealed it would have been nearly restored to the strength of the drilled one. If the holes had been punched $\frac{1}{8}''$ small in diameter and reamed to size, the plate would have been as strong as the drilled one. These facts, which have been experimentally determined, point to the following conclusions: First, punching injures the material and produces weakness. Second, the injury is due to stresses caused by the severe action of the punch, since annealing, which furnishes opportunity for equalization of stress, restores the strength. Third, the injury is only in the immediate vicinity of the punched hole, since reaming out $\frac{1}{16}''$ on a side removes all the injured material. In ordinary boiler work the plates are simply punched and riveted. If better work is required the plates must be drilled, or punched small and reamed, or punched and annealed. Drilling is slow and therefore expensive; annealing is apt to change the plates and requires large expensive furnaces. Punching small and reaming, is probably the best method.

95. **Riveted Joints** are of two general kinds: First, **Lap Joints**, in which the sheets to be joined are lapped upon each other and joined by a seam of rivets, as in Fig. 104 *a*. Second, **Butt Joints**, in which the edges of the sheets abut against each other, and a strip called a "cover plate" or "butt strap" is riveted to both edges of the sheet, as in *c*.

There are two kinds of riveting: Single, in which there is but one row of rivets, as in *a*; and double, where there are two rows.

Double riveting is subdivided into "chain riveting," *b*, and "zigzag" or "staggered" riveting, *d*.

Lap joints may be single, double chain, or double staggered riveted.

Butt joints may have a single strap, as in *c*, or double strap; *i. e.*, an exactly similar one is placed on the other side of the joint. Butt joints with either single or double strap may be single, double chain, or double staggered riveted.

To sum up, there are:

Lap Joints { Single   Riveted
             Double Chain   "
               "   Staggered "

Butt Joints { Single Strap { Single   Riveted
                             Double Chain   "
                               "   Staggered "
              Double   " { Single   Riveted
                             Double Chain   "
                               "   Staggered "

A riveted joint may yield in any one of four ways: First, by the rivet shearing. Second, by the plate yielding to tension on the line AB, Fig. 105 *a*. Third, by the rivet tearing out through the margin, as in *c*. Fourth, the rivet and sheet bear upon each other at D and E in *d*, and are both in compression. If the unit stress upon these surfaces becomes too great, the rivet is weakened to resist shearing, or the plate to resist tension, and failure may occur. This pressure of the rivet on the sheet is called "bearing pressure."

96. As a preliminary to the designing of joints it is necessary to know the strength of the rivets to resist shear; of the plate to resist tension; and of the rivets and plate to resist bearing pressure. These values must not be taken from tables of the strength of the materials of which the plate and rivets are made, but must be derived from experiments upon actual riveted joints tested to rupture. The reason for this is that the conditions of stress are modi-

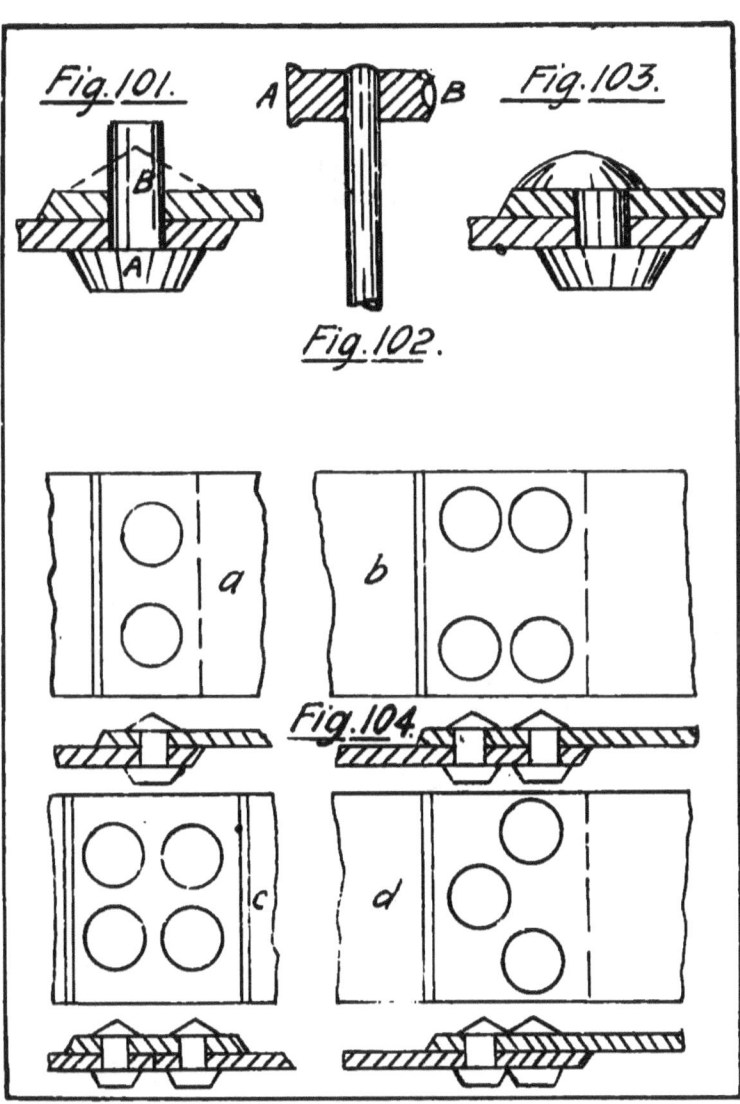

fied somewhat in the joint. For instance, in single strap butt joints, and in lap joints, the line of stress being the centre line of plates, and the plates joined being offset, flexure results and the plate is weaker to resist tension; if the joint yield to this stress in the slightest degree the "bearing pressure" is localized, and becomes more destructive. Extensive and accurate experiments have been made upon actual joints and the results are available in Stoney's "Strength and Proportions of Riveted Joints." The constants given are taken from this book.

	Iron.	Steel.
Ultimate shearing strength of rivets, single shear	40000	·50000
" " " " double "	35000	44000
Ultimate tensile strength of plate between rivet holes, single shear	40000	60000
Ultimate bearing pressure per sq. inch of diametral plane of rivet, single shear	67000	95000
Ultimate bearing pressure per sq. inch of diametral plane of rivet, double shear........................	89000	100000

97. The theoretical diameter of rivet for a given thickness of plate may now be determined. Let $d =$ diameter of rivet hole; $t =$ thickness of plate; $p =$ pitch of rivets; $T =$ ultimate tensile strength of plate between rivet holes; $S =$ ultimate shearing strength of rivets; $C =$ ultimate bearing pressure.

The strength of the rivet to resist shearing at AB, Fig. 106, should be equal to its strength to resist bearing pressure at A and C, and hence the expressions for those strengths may be equated, thus:

$$Ctd = Sd^2 \frac{\pi}{4}.$$

Solving, $$d = \frac{Ct}{S \times 0.7854} = \frac{67000}{40000 \times 0.7854} t = 2t.$$

Hence, for equal strength to resist bearing pressure and shear, the diameter of the rivet should equal twice the thickness of the plate.

Let the results thus derived be compared with the values that are used in actual practice. See table.

COMPARATIVE VALUES IN INCHES OF RIVET DIAMETER FOR DIFFERENT VALUES OF THICKNESS OF PLATE.

t	$2t$	$1.2\sqrt{t}$	d
$3/16$	$3/8$	—	$3/8$
$1/4$	$1/2$	$9/16$	$1/2$
$5/16$	$5/8$	$11/16$	$5/8$
$3/8$	$3/4$	$3/4$	$5/8$–$3/4$
$1/2$	1	$7/8$	$3/4$–$7/8$
$5/8$	$1\,1/4$	$15/16$	$3/4$–1
$3/4$	$1\,1/2$	$1\,1/16$	1–$1\,1/8$
$7/8$	$1\,3/4$	$1\,1/8$	1–$1\,3/8$
1	2	$1\,3/16$	1–$1\,1/4$
$1\,1/8$	$2\,1/4$	—	$1\,1/8$–$1\,3/8$

The first column gives the thickness of the plate; the second the diameter of the rivet $= 2t$; the third gives the rivet diameter calculated from the formula of Professor Unwin, $d = 1.2\sqrt{t}$; the fourth column gives rivet diameters as found in practice, taken from Stoney's book, page 12. $d = 2t$ agrees with practice up to $3/8''$ plates, but for thicker plates it gives values that are too large. The reason for this is that the difficulty in driving rivets increases very rapidly with their size; $1\,1/4''$ or $1\,3/8''$ being the largest rivet that can be driven conveniently. The equality of strength to resist bearing pressure and shear is therefore sacrificed to convenience in manipulation. As the diameter of the rivet is increased the area to resist bearing pressure increases less rapidly than the area to resist shear (the thickness of the plate remaining the same), the former varying as

d, and the latter as d^2; therefore if d is not increased as much as is necessary for equality of strength, the excess of strength will be to resist bearing pressure. If the other parts of the joint are made as strong as the rivet in shear, and this strength is calculated from the stress to be resisted, the joint will evidently be correctly proportioned.

To calculate the diameter of rivet for a butt joint with double cover plates. — The rivet is in double shear, and therefore ultimate bearing pressure $= 89000$ lbs. per square inch $= C$. And also ultimate shear pressure $= 35000$ lbs. per square inch $= S'$.

Equating as before $\quad Cdt = \dfrac{S'\pi d^2 2}{4} = \dfrac{S'\pi d^2}{2}$.

From which $\quad d = \dfrac{2Ct}{S'\pi} = \dfrac{2 \times 89000 \times t}{\pi \times 35000} = 1\cdot 6\,t$ nearly.

Comparison of results of this formula with tables of dimensions of practice, shows them to be too large. The following empirical formulas may be trusted:

For thin plates — for iron $d = 1\cdot 3\,t$; for steel, $d = 1\cdot 25\,t$.
" thick " " $d = 1\cdot 1\,t$; " $d = 1\cdot 125\,t$.

98. The next value to be determined is the *pitch* of the rivets, *i. e.*, the distance from the centre of one rivet to the centre of the next one. See Fig. 107. It is required to make the pitch of such a value that the strength of the plate between rivet holes to resist tension shall equal the strength of the rivet to resist shear. It has already been shown that the strength to resist bearing pressure is equal to, or greater than, the strength to resist shear. Equate expressions for shearing strength of the rivet, and tensile strength of the plate on a section through the rivet holes, and solve for $p =$ *pitch*. For a single riveted lap joint,

$$\dfrac{\pi d^2}{4} S = Tt\,(p - d).$$

From which $\quad p = \dfrac{0\cdot 7854 d^2 S + Ttd}{Tt}$.

106   MACHINE DESIGN.

Let $S = 40000$ and $T = 40000$.
Then if $t = \frac{1}{4}''$, $d = \frac{1}{2}''$; $p = 1.28''$.
$t = \frac{3}{8}''$, $d = \frac{3}{4}''$; $p = 1.79''$.
$t = \frac{1}{2}''$, $d = \frac{7}{8}''$; $p = 2.06''$.
$t = 1''$, $d = 1\frac{1}{8}''$; $p = 2.12''$.

All of these agree with Stoney's "Table of Boilermaker's Proportions," lap joints, iron plates, and rivets, except for $t = \frac{1}{4}''$. This formula may, therefore, be trusted except for very thin plates.

To find p for butt joints with double straps, single riveted.—Since the rivet is in double shear,

$$p = \frac{2 \times 0\cdot7854 d^2 S' + Ttd}{Tt};$$

$S' = 35000$ lbs. per square inch, the value for double shear.

For steel plates and steel rivets, the values of the constants, T and S, need to be changed in above formulas. See values given.

To find the pitch of double riveted joints the method is the same. There are, however, two rivets now to support the strip of plate between holes, instead of one, as in the single joint. See Fig. 107. Therefore the first formula for p, multiplying the shearing strength by 2, becomes

$$p = \frac{1\cdot57 d^2 S + Ttd}{Tt}.$$

For double shear $$p = \frac{3\cdot14 d^2 S' + Ttd}{Tt};$$

S' being value for double shear.

99. The *margin* in a riveted joint is the distance from the edge of the sheet to the rivet hole. This must be made of such value that there shall be safety against failure by the rivet tearing out. There can be no satisfactory theoretical determination of this value; but practice and experiments with actual joints show that a joint will not yield in this way if the margin be made $= d =$ diameter of the rivet. The distance between the centre lines of the rows in double chain riveting may be taken $= 2\cdot5 d$; and in double stag-

RIVETED JOINTS.   107

gered riveting may be taken $= 1\cdot 88d$. Thus the total width of lap for single riveting equals $3d$; in double chain riveting $= 5\cdot 5d$; and in double staggered riveting $4\cdot 88d$. The riveted joints considered cannot be as strong as the unperforated plates. The ratio of strength of joint to strength of plate is called *joint efficiency*. If the joint were equally strong to resist rupture in all possible ways, the joint efficiency would equal the ratio of area of plate through rivet section, to the area of unperforated section. Results obtained in this way differ somewhat from the results of actual tests, and the latter values should be used. See following tables.

RELATIVE EFFICIENCY OF IRON JOINTS.

	Efficiency Per Cent.
Original solid plate	100
Lap Joint, single riveted, punched	45
" " " drilled	50
" ·double "	60
Butt Joint, single cover, single riveted	45–50
" " " double "	60
" double " single "	55
" " " double "	66

RELATIVE EFFICIENCY OF STEEL JOINTS.

	Efficiency Per Cent.		
	Thickness of Plates.		
	¼–⅜	½–⅝	¾–⅞
Original solid plate	100	100	100
Lap Joint, single riveted, punched	50	45	40
" " " drilled	55	50	45
" double " punched	75	70	65
" " " drilled	80	75	70
Butt Joint, double cover, double riveted, punched	75	70	65
" " " " drilled	80	75	70

These tables are from Stoney's "Strength and Proportions of Riveted Joints."

108   MACHINE DESIGN.

100. The following problem will serve to illustrate the design of riveted joints for boilers. It is required to design a horizontal tubular boiler 48" diameter to carry a working pressure of 100 lbs. per square inch. A boiler of this type consists of a cylindrical shell of wrought iron or steel plates made up in length of two or more courses or sections. Each course is made by rolling a flat sheet into a hollow cylinder and joining its edges by means of a riveted joint, called the longitudinal joint or seam. The courses are joined to each other also by riveted joints, called circular joints or seams. Circular heads of the same material have a flange turned all around their circumference, by means of which they are riveted to the shell. The proper thickness of plate may be determined from: I. The diameter of shell $= 48''$. II. The working steam pressure per square inch $= 100$ lbs. III. The tensile strength of the material used; let steel plates be used of 60000 lbs. specified tensile strength. Preliminary investigation of the conditions of stress in the cross-section of material cut by a plane.—I. Through the axis; II. At right angles to the axis, of a thin hollow cylinder; the stress being due to the excess of internal pressure per square inch over the external pressure per square inch. Let $l =$ the length of the cylindrical shell in inches; $d =$ the diameter of the cylindrical shell in inches; $p =$ the excess of internal over external pressure per square inch; $p_1 =$ unit stress in a longitudinal section of material of the shell due to p; $p_2 =$ unit stress in a circular section of material of the shell due to p; $t =$ thickness of plate; $T =$ ultimate tensile strength of plate.

In a longitudinal section the stress $= ldp$, and the area of metal sustaining it $= 2lt$. Then $p_1 = \dfrac{dp}{2t}$.

In a circular section the stress $= \dfrac{\pi d^2 p}{4}$, and the area $= \pi d t$ nearly. Then $p_2 = \dfrac{\pi d^2 p}{4} \times \dfrac{1}{\pi d t} = \dfrac{dp}{4t}$.

Therefore the stress in the first case is twice as great as in the second; and a thin hollow cylinder is twice as strong to resist rupture on a circular section as on a longitudinal one. The latter only, therefore, need be considered in determining the thickness of plate. Equating the stress due to p in a longitudinal section and the strength of the cross-section of plate that sustains it, we have $ldp = 2ltT$. Therefore $t = \dfrac{dp}{2T} =$ the thickness of plate that would just yield to the unit pressure p. To get safe thickness, a factor of safety must be used. It is usually equal in boiler shells to 4, 5, or 6. Its value is small because the material is highly resilient and the changes of pressure are gradual, *i. e.*, there are no shocks. This takes no account of the riveted joint, which is the weakest longitudinal section, e times as strong as the solid plate; e being the joint *efficiency*, $= 0.75$ if the joint be double riveted. The formula then becomes $t = \dfrac{fdp}{2Te}$. Substituting values

$$t = \frac{6 \times 48 \times 100}{2 \times 60000 \times 0.75} = 0.32'', \text{ say } {}^{5}\!/_{16}''.$$

The circular joints will be single riveted and joint efficiency will $= 0.50$. But the stress is only one-half as great as in the longitudinal joint, and therefore it is stronger in the proportion 0.50×2 to $0.75 = 1$ to 0.75. From this it is seen that a circular joint whose efficiency is 0.50 is as strong as the solid plate in a longitudinal section. From the value of t the joints may now be designed. Diameter of rivet $= d = 1.2\sqrt{t} = 1.2\sqrt{0.3125} = 0.672''$, say $0.687'' = {}^{11}\!/_{16}''$. The pitch for a single riveted joint $=$

$$p = \frac{0.7854 d^2 S + Ttd}{Tt};$$

But $d = {}^{11}\!/_{16} = .687''$; $S = 50000$ for steel; $T = 60000$ for steel; $t = {}^{5}\!/_{16} = 0.3125$. Substituting these values $p = 1.42''$. For double riveted joint

$$p = \frac{1.57 d^2 S + Ttd}{Tt} = \text{(substituting as above) } 2.66''.$$

The margin $= d = 0.687'' = {}^{11}/_{16}''$. The longitudinal joint will be staggered riveted and the distance between rows $= 1.88d = 1.29'' =$ say $1\,{}^{5}/_{16}''$. The total lap in the longitudinal joint $= 4.88d = 3.35''$. The total lap in the circular joint $= 3d = 2\,{}^{1}/_{16}''$. The joints are therefore completely determined, and a detail of each, giving dimensions, may be drawn for the use of the workmen who make the templets and lay out the sheets.

CHAPTER X.

DESIGN OF JOURNALS.

101. Journals and the bearings or boxes with which they engage are the elements used to constrain motion of rotation or vibration about axes in machines. Journals are usually cylindrical, but may be conical, or, in rare cases, spherical. The design of journals, as far as size is concerned, is dictated by one or all of the three following considerations: I. To provide for safety against rupture or excessive yielding under the applied forces. II. To provide for maintenance of form. III. To provide against the squeezing out of the lubricant. To illustrate I. — Let Fig. 108 represent a pulley on the end of an overhanging shaft, driven by a belt, ABC. Rotation is as indicated by the arrow, and the belt tensions are T_1 and T_2. The journal, J, engages with a box or bearing, D. The following stresses are induced in the journal: *Torsion*, measured by the torsional moment $(T_1 - T_2)r$. *Flexure*, measured by the bending moment $(T_1 + T_2)a$. This assumes a rigid shaft or a self-adjusting box. *Shear*, resulting from the force $T_1 + T_2$. This journal must therefore be so designed that rupture or undue yielding shall not result from any one of these stresses. To illustrate II. — Consider the spindle journals of a grinding lathe. The forces applied are very small; but the *form* of the journals must be maintained to insure accuracy in the product of the machine. A relatively large wearing surface is therefore necessary. To illustrate III. — The pressure upon a journal resulting from the applied forces may be sufficiently great to squeeze out the lubricant. Metallic contact, heating, and abrasion of the surfaces would result. In what follows, the area of a journal means its *projected* area ; *i. e.*, its length × its diameter.

The allowable pressure per square inch of area of a journal varies with several conditions. To make this clear, suppose a drop of oil to be put in the middle of an accurately finished surface plate; suppose another exactly similar plate to be placed upon it for an instant; the oil drop will be spread out because of the force due to the weight of the upper plate. If the plate were allowed to remain a longer time, the oil would be still further spread out, and if its weight and the time were sufficient, the oil would finally be entirely squeezed out from between the plates, and the metal surfaces would come in contact. The squeezing out of the oil from between the rubbing surfaces of a journal and its box is, therefore, a function of the *time* as well as of pressure. If the surfaces under pressure move over each other, the removal of the oil is facilitated. The greater the velocity of movement, the more rapidly will the oil be removed, and therefore the squeezing out of the oil is also a function of the *velocity of rubbing surfaces.*

When a journal is subjected to continuous pressure in one direction, as for instance in a shaft with a constant belt pull, or with a heavy fly-wheel upon it, this pressure has sufficient time to act, and is therefore effective for the removal of the oil. But if the direction of the pressure is periodically reversed, as in the crank-pin of a steam engine, the time of action is less, the tendency to remove the oil is reduced, and the oil has opportunity to return between the surfaces. Hence, a higher pressure per square inch of journal would be allowable in the second case than in the first.

If the direction of motion is also reversed, as in the cross head pin of a steam engine, the oil not only has an opportunity to return between the surfaces, but is assisted in doing so by the reversed motion. Therefore, a still higher pressure per square inch of journal is allowable. Practical experience bears out these conclusions. Thus in journals with the direction of pressure constant, it is found that with ordinary conditions of lubrication the heating and "seizing" or "cutting" occur quickly if the pressure per square inch of

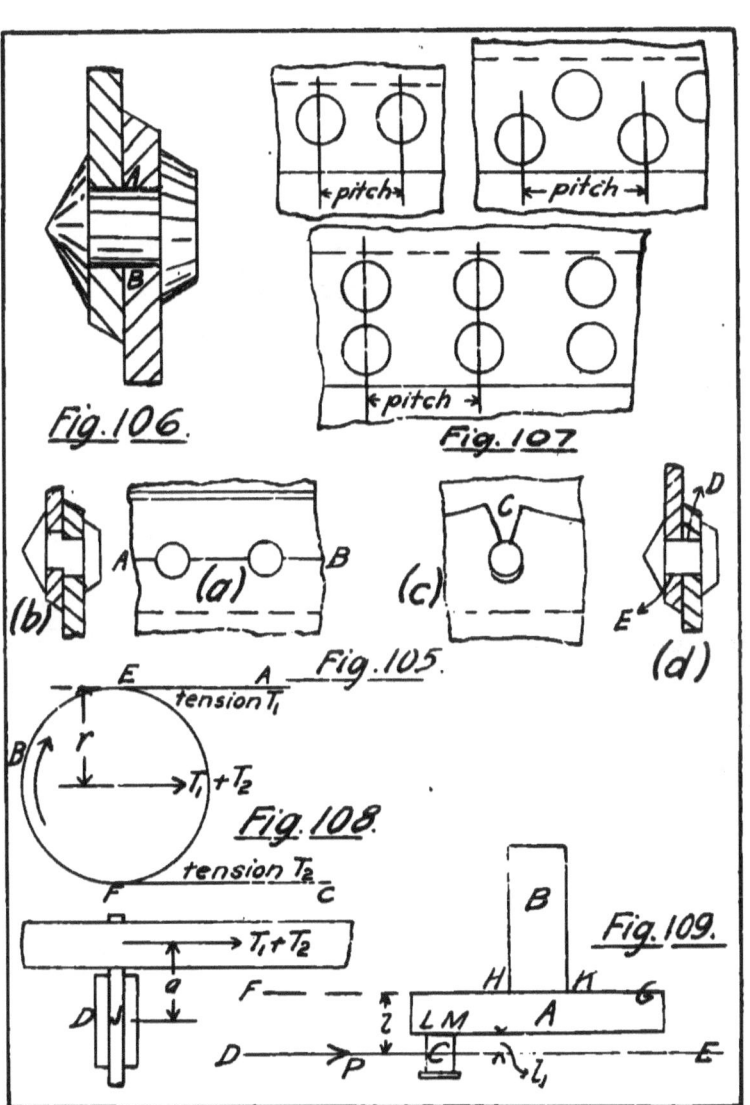

journal exceed about 380 lbs.* But in the crank-pins of punching machines, where the pressure acts for an instant, with quite an interval of rest, and where the velocity of rubbing surface is very low indeed, the pressure is often as high as from 2000 to 3000 lbs., per square inch, and there is no tendency to heating or abrasion. In engine crank-pins the pressure may be from 400 to 800 lbs. depending on the velocity of rubbing surface, and in cross-head pins where the velocity is always low it may be from 600 to 1000 lbs. The value to be used in each particular case must be decided by the judgment of the designer.

But even if the conditions are such that the lubricant is retained between the rubbing surfaces, heating may occur. There is always a frictional resistance at the surface of the journal; this resistance may be reduced: *a*, by insuring accuracy of form and perfection of surface in the journal and its bearings; *b*, by insuring that the journal and its bearing are in contact, except for the film of oil, throughout their entire surface, by means of rigidity of framing or self-adjusting boxes, as the case may demand; *c*, by selecting a suitable lubricant to meet the conditions, and maintaining the supply to the bearing surfaces. By these means the friction may be reduced to a very low value, but it cannot be reduced to zero.

There must be some frictional resistance, and it is always converting mechanical energy into heat. This heat raises the temperature of the journal and its bearing. If the heat thus generated is conducted and radiated away as fast as it is generated, the box remains at a constant low temperature. If, however, the heat is generated faster than it can be disposed of, the temperature of the box rises till its capacity to radiate heat is increased by the increased difference of temperature of the box and the surrounding air, so that it is able to dispose of the heat as fast as it is generated. This temperature, necessary to establish the equilibrium of heat generation and disposal, may under certain conditions be high enough to destroy the lubricant, or even to melt out a babbitt metal

* See Mr. Tower's experiments in the "Minutes of the Institution of Mechanical Engineers."

box lining. Suppose now that a journal is running under certain conditions of pressure and surface velocity, and that it remains entirely cool. Suppose next that while all other conditions are kept exactly the same, the velocity is increased. All modern experiments on the friction in journals show that the friction increases with the increase of the velocity of rubbing surface. Therefore the increase in velocity would increase the frictional resistance at the surface of the journal, and the space through which this resistance acts would be greater in proportion to the increase in velocity. The work of the friction at the surface of the journal is therefore increased, because both the force and the space factors are increased. It is this work of friction which has been so increased, that produces the heat that tends to raise the temperature of the journal and its box. The rate of generation of heat has therefore been increased by the increase in velocity, but the box has not been changed in any way and therefore its capacity for disposing of heat is the same as it was before, and hence the tendency of the journal and its bearing to heat is greater than it was before the increase in velocity. Some change in the proportions of the journal must be made in order to keep the tendency to heat the same as it was before the increase in velocity. If the diameter of the journal be increased, the radiating surface of the box will be proportionately increased. But the space factor of the friction will be increased in the same proportion, and therefore it will be apparent that this change has not affected the relation of the rate of generation of heat to the disposal of it. But if the length of the journal be increased, the work of friction is the same as before and the radiating surface of the box is increased and the tendency of the box to heat is reduced. If, therefore, the conditions are such that the tendency to heat in a journal, because of the work of the friction at its surface, is the vital point in design, it will be clear that the length of the journal is dictated by it, but not the diameter. The reason why high speed journals have greater length in proportion to their diameter than low speed journals will now be apparent.

102. *Problem.*—To design the main journal of a side crank

DESIGN OF JOURNALS. 115

engine. — The data are as follows: Diameter of steam cylinder = 16"; boiler pressure = 100 lbs. per sq. in. by gauge. Then the maximum force upon the steam piston due to steam pressure = $100 \times 8^2\pi = 20106$ lbs. Suppose that the least expensive stress member is a "breaking piece," *i. e.*, it will yield and relieve the stress in the other stress members when the total applied force = 80000 lbs., about four times the maximum working force. In Fig. 109, DE is the centre line of the engine; C is the crank-pin; A is the crank disc, and B is the journal to be designed. The force P, = 80000 lbs., is the greatest force that can act in the line DE. The journal is supported up to the line FG. In the section HK there is flexure stress measured by the flexure moment Pl. l in this case = 6". The breaking piece only yields when the crank is at or near its centre; hence, the torsional stress may be neglected. The radius of the shaft for safety against the moment Pl, may be found from the formula

$$Pl = \frac{SI}{c}; \text{ from which } I = \frac{Plc}{S}. \text{ But } I \text{ for circular section} = \frac{\pi r^4}{4};$$

and $c = r$. Hence $r^3 = \frac{4Pl}{\pi S}$. Let S for machinery steel = 12000. This gives a factor of safety $= \frac{60000}{12000} = 5$. Substituting values, $P = 80000$, $l = 6"$, and $S = 12000$, in the above equation, gives $r = 3.71"$, say $3\frac{3}{4}"$. Hence, the shaft diameter = $7\frac{1}{2}"$. This value depends upon the assumptions made for P, the strength of the breaking piece, and for S, the safe stress for the material used. Different values might have been assumed, and would, of course, have given different results. The length of such a journal is determined by practical considerations. In this case the length should be about twice the diameter = 15", in order that convenient means may be supplied for taking up wear. The projected area of journal = $7.5" \times 15 = 112.5$ square inches. Assume 350 lbs. safe pressure per square inch of journal. This would admit of a working pressure of $350 \times 112.5 = 39375$. It is evident without investigation that this is greater than any working load for this journal.

116   MACHINE DESIGN.

103. *Problem*. — To design the crank-pin for the same engine. — The bending moment now equals Pl_1. Assume $l_1 = 3''$. Then $r^3 = \dfrac{4 \times 80000 \times 3}{\pi \times 12000}$, from which $r = 2\cdot 94$, say $3''$. Therefore, $d = 6''$; and since the assumed length $= 6''$, the journal area $= 36$ sq. in. Then if the allowable pressure per square inch $= 700$ lbs., the *total* allowable working pressure $= 25200$ lbs. This is greater than the possible working pressure, and hence the lubricant would not be squeezed out. The size of both journal and crank-pin is therefore dictated by the maximum bending moment.

104. To design the cross-head pin for the same engine. — In Fig. 110, C represents the cross-head pin. The force P, $= 80000$, may be applied as indicated. The pin is supported at both ends, and the connecting-rod box bears upon it throughout its entire length, AD or BE. The pin would yield by shearing on the sections AB and DE. The shearing strength of the machinery steel, of which it would be made, may be assumed equal to 50000 lbs. A stress of 8000 lbs. would therefore give a factor of safety of $6 +$. The necessary area in shear would equal $\dfrac{80000}{8000} = 10$, or 5 square inches for each section. This corresponds to a diameter of $2\cdot 5 +$. The length of pin may be found as follows: Find the mean working force upon the pin, by drawing the indicator card, as modified by acceleration of reciprocating parts, and multiplying the value of its mean ordinate, in pounds per square inch, by the piston area. The value for this case $=$ about 12000 lbs. The allowable pressure per square inch of journal $= 800$ lbs. Hence the journal area $= 12000 \div 800 = 15$. The length then $= 15 \div 2\cdot 5 = 6''$. The diameter of the cross-head pin, therefore, is dictated by the applied force, while its length depends upon the maintenance of lubrication. The judgment of the designer might require this pin to be still larger to reduce wear and to maintain its form.

Journals whose maintenance of form is of chief importance, must be designed from precedent, or according to the judgment of the designer. No theory can lead to correct proportions. In fact these proportions are eventually determined by the process of Machine Evolution.

105. Thrust Journals. — When a rotating machine part is subjected to pressure parallel to the axis of rotation, means must be provided for the safe resistance of that pressure. In the case of vertical shafts the pressure is due to the weight of the shaft and its attached parts — as the shafts of turbine water-wheels that rotate about vertical axes. In other cases the pressure is due to the working force — as the shafts of propeller wheels, the spindles of a chucking lathe, etc. The end thrust of vertical shafts is very often resisted by the "squared up" end of the shaft. This is inserted in a bronze or brass "bush," which embraces it to prevent lateral motion, as in Fig. 111. If the pressure be too great, the end of the shaft may be enlarged so as to increase the bearing surface, thereby reducing the pressure per square inch. This enlargement must be within narrow limits, however. See Fig. 112. AB is the axis of rotation, and ACD is the rotating part, its bearing being enlarged at CD. Let the conditions of wear be considered. The velocity of rubbing surface varies from zero at the axis to a maximum at C and D. It has been seen that the increase of the velocity of rubbing surface increases both the force of the friction and the space through which that force acts; it therefore increases the work of the friction, and therefore the tendency to wear. From this it will be seen that the tendency to wear increases from the centre to the circumference of this "radial bearing," and that, after the bearing has run for a while, the pressure will be localized near the centre, and heating and abrasion may result. For this reason, where there is severe stress to be resisted, the bearing is usually divided up into several parts, the result being what is known as a "collar thrust bearing," as shown in Fig. 113. By the increase in the number of collars, the bearing surface may be increased without increasing the tendency to unequal wear. The radial dimension of the bearing is kept as small as is consistent with the other considerations of the design. It is found that the "tractrix," the curve of constant tangent, gives the same work of friction, and hence the same tendency to wear in the direction of the axis of rotation, for all parts of the wearing surface. (See "Church's Mechanics," page 181.) This is without doubt the best form for a thrust bearing, but the difficulties in the way of the accurate production of its curved outlines have interfered with its extensive use.

The pressure that is allowable per square inch of projected area of the bearing surface varies in thrust bearings with several conditions, as it does in journals subjected to pressure at right angles to the axis. Thus in the pivots of turn-tables, swing bridges, cranes, and the like, the movement is slow and never continuous, often being reversed; and also the conditions are such that "bath lubrication" may be used, and the allowable unit pressure is very high — equal often to 1500 pounds per square inch, and in some cases greatly exceeding that value. The following table may be used as an approximate guide in the designing of thrust bearings. The material of the thrust journal is wrought iron or steel, and the bearing is of bronze or brass (babbitt metal is seldom used for this purpose). Bath lubrication is used, $i.\ e.$, the running surfaces are submerged constantly in a bath of oil.

Mean Velocity of rubbing surface, feet per minute.	Allowable Unit Pressure, lbs. per square inch of projected area of the rubbing surface.
Up to 50	1000
50 to 100	600
100 to 150	350
150 to 200	100
Above 200	50

If the journal is of cast iron and runs on bronze or brass, the values of allowable pressure given should be divided by 2.

106. Examples to illustrate the design of thrust journals.

Example I. — It is required to design a thrust journal whose outline is a tractrix. It is required to support a vertical shaft which with its attached parts weighs 2000 lbs., and runs at a rotative speed of 200 revolutions per minute. The dimensions of the thrust journal are

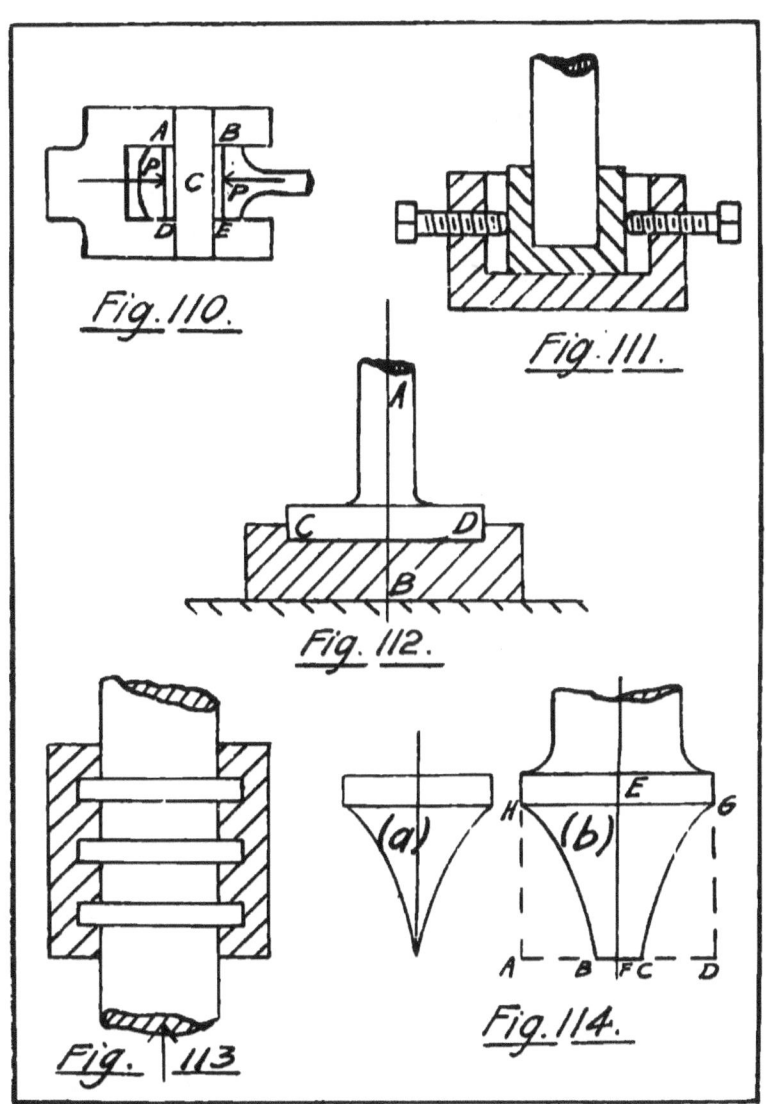

DESIGN OF JOURNALS.   119

as yet unknown, and therefore the velocity of rubbing surface must be estimated. Suppose that the mean diameter of the journal is 2"; then the mean velocity of rubbing surface will be $2 \times \pi \times N \div 12 = 103$ feet per minute. This is so near the limit in the table between an allowable pressure of 350 and 600 that an intermediate value may be used, say 450 pounds. The projected area of the journal then will equal the total pressure divided by the allowable pressure per square inch of the journal $= 2000 \div 450 = 4\cdot44$ square inches. The journal must not be pointed, as in (a) Fig. 114, but must be as shown in (b). The dimension BC may be assumed equal to 1". The projected area of the journal is equal to the circular area whose diameter is AD, minus the circular area whose diameter is BC, and this may be equated with the required value, equal 4·44, and the equation solved for the required dimension, AD.

Thus $$\frac{(AD)^2\pi}{4} - \frac{(BC)^2\pi}{4} = 4\cdot 44$$

Therefore $$(AD)^2 = \frac{4\cdot 44 \times 4}{\pi} + (BC)^2.$$

$$AD = \sqrt{6\cdot 68} = 2\cdot 58".$$

In order now to draw the required journal, lay off from the axis EF the distance EG, equal half AD, and through the point G draw a tractrix whose constant tangent is equal to EG, continuing the curve till it reaches a point C, such that FC is equal to half the assumed value of BC. The vertical dimension of the journal is thereby determined, and the corresponding curve, BH, may be drawn on the other side of the axis EF.

107. *Example II.*— It is required to design the collar thrust journal that is to receive the propelling pressure from the screw of a small yacht. The necessary data are as follows: The maximum power delivered to the shaft is 70 H. P.; pitch of screw is 4 feet: slip of screw is 20%; shaft revolves 250 times per minute; diameter of shaft is 4".

For every revolution of the screw the yacht moves forward a

distance = 4 ft. less 20% = 3·2 ft., and the speed of the yacht in feet per minute = 250 × 3·2 = 800. 70 H. P. = 70 × 33000 = 2,310,000 foot-pounds per minute. This work may be resolved into its factors of force and space, and the propelling force is equal to 2,310,000 ÷ 800 = 2900 lbs. nearly. The shaft is 4" diameter, and the collars must project beyond its surface. Estimate that the mean radius of the rubbing surface is 4·5", then the mean velocity of rubbing surface would equal 4·5 × π ÷ 12 × 250 = 294 feet per minute. The allowable value of pressure per square inch of journal surface for a velocity above 200 ft. per minute is 50 lbs. The necessary area of the journal surface is therefore = 2900 ÷ 50 = 58 square inches. It has been seen that it is desirable to keep the radial dimension of the collar surface as small as possible in order to have as nearly the same velocity at all parts of the rubbing surface as possible. The width of collar in this case will be assumed = 0·75"; then the bearing surface in each collar

$$= \frac{5·5^2 \times \pi}{4} - \frac{4^2 \times \pi}{4} = 23·7 - 12·5 = 11·2.$$

Then the number of collars equals the total required area divided by the area of each collar = 58 ÷ 11·2 = 5·18, say 6.

108. Bearings and Boxes. — The function of a bearing or box is to insure that the journal with which it engages shall have an accurate motion of rotation or vibration about the given axis. It must therefore fit the journal without lost motion; must afford means of taking up the lost motion that results necessarily from wear; must resist the forces that come upon it through the journal, without undue yielding; must have the wearing surface of such material as will run in contact with the material of the journal with the least possible friction, and least tendency to heating and abrasion; and must usually include some device for the maintenance of the lubrication. The selection of the materials and the providing of sufficient strength and stiffness depends upon principles already considered, and so it remains to discuss the means for the taking up of necessary wear and for providing lubrication.

DESIGN OF JOURNALS.   121

Boxes are sometimes made solid rings or shells, the journal being inserted endwise. In this case the wear can only be taken up by making the engaging surfaces of the box and journal conical, and providing for endwise adjustment either of the box itself or of the part carrying the journal. Thus, in Fig. 115, the collars for the preventing of end motion while running, are jamb nuts, and looseness between the journal and box may be taken up by moving the journal axially toward the left.

By far the greater number of boxes, however, are made in sections, and the lost motion is taken up by moving one or more sections toward the axis of rotation. The tendency to wear is usually in one direction, and it is sufficient to divide the box into halves. Thus, in Fig. 116, the journal rotates about the axis O, and all the wear is due to the pressure P acting in the direction shown. The wear will therefore be at the bottom of the box. It will suffice for the taking up of wear to dress off the surfaces at aa, and thus the box cap may be drawn further down by the bolts, and the lost motion is reduced to an admissible value. "Liners" or "shims," which are thin pieces of sheet metal, may be inserted between the surfaces of division of the box at aa, and may be removed successively for the lowering of the box cap as the wear renders it necessary. If the axis of the journal must be kept in a constant position, the lower half of the box must be capable of being raised.

Sometimes, as in the case of the box for the main journal of a steam engine shaft, the direction of wear is not constant. Thus, in Fig. 117, A represents the main shaft of an engine. There is a tendency to wear in the direction B because of the weight of the shaft and its attached parts; there is also a tendency to wear because of the pressure that comes through the connecting-rod and crank. The direction of this pressure is constantly varying, but the average direction on forward and return stroke may be represented by C and D. Provision needs to be made, therefore, for taking up wear in these two directions. If the box be divided on the line EF, wear will be taken up vertically and horizontally by reducing the liners. Usually, however, in the larger engines the box is divided into four sections, A, B, C, and D (Fig. 118), and A and C are capable of being moved toward the shaft by means of screws or wedges, while D may be raised by the insertion of "shims."

The lost motion between a journal and its box is sometimes taken up by making the box as shown in Fig. 119. The external surface of the box is conical and fits in a conical hole in the machine frame. The box is split entirely through at A, parallel to the axis, and partly through at B and C. The ends of the box are threaded, and the nuts E and F are screwed on. After the journal has run long enough so that there is an unallowable amount of lost motion, the nut F is loosened and E is screwed up; the effect being to draw the conical box further into the conical hole in the machine frame; the hole through the box is thereby closed up, and lost motion is reduced. After this operation the hole cannot be truly cylindrical, and if the cylindrical form of the journal has been maintained, it will not have a bearing throughout its entire surface. This is not usually of very great importance, however, and the form of box has the advantage that it holds the axis of the journal in a constant position.

All boxes in self-contained machines, like engines or machine tools, need to be rigidly supported to prevent the localization of pressure, since the parts that carry the journals are made as rigid as possible. In line shafts and other parts carrying journals, when the length is great in comparison to the lateral dimensions, some yielding must necessarily occur, and if the boxes were rigid, localization of pressure would result. Hence "self-adjusting boxes" are used. A point in the axis of rotation at the centre of the length of the box is held immovable, but the box is free to move in any way about this point, and thus adjusts itself to any yielding of the shaft. This result is attained as shown in Fig. 120. O is the centre of the motion of the box; B and A are spherical surfaces formed on the box, their centre being at O. The support for the box carries internal spherical surfaces which engage with A and B.

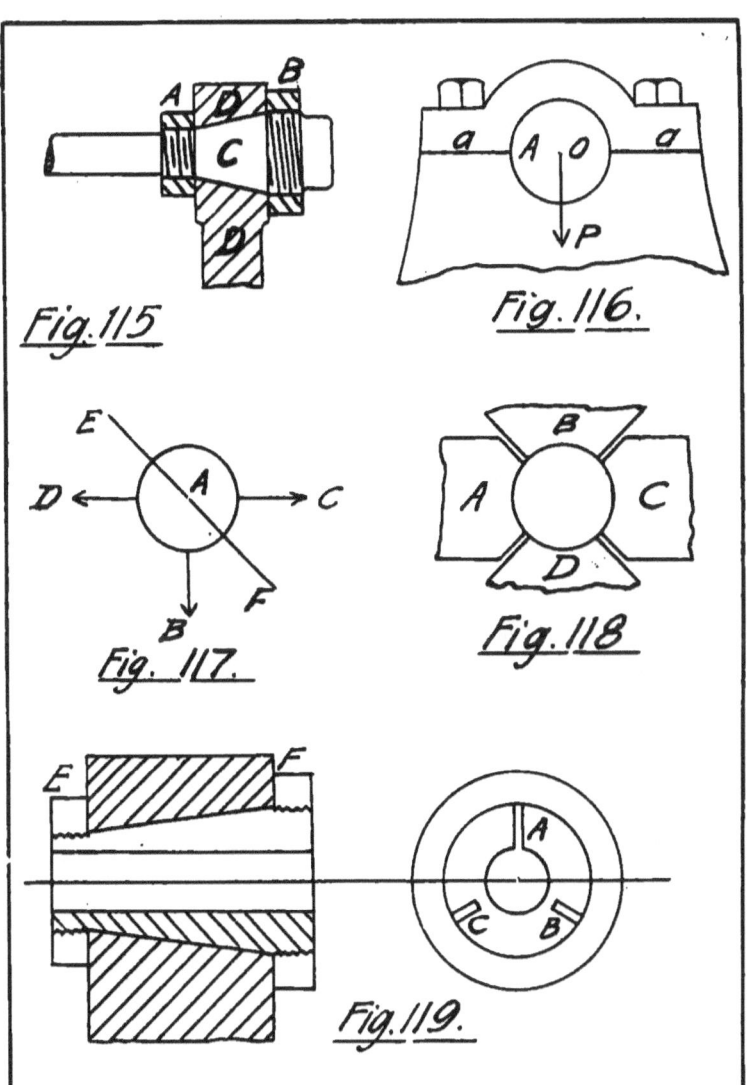

Thus the point O is always held in a constant position, but the box itself is free to move in any way about O as a centre. Therefore the box adjusts itself, within limits, to any position of the shaft, and hence the localization of pressure is impossible.

In thrust bearings for vertical shafts the weight of the shaft and its attached parts serves to hold the rubbing surfaces in contact, and the lost motion is taken up by the shaft following down as wear occurs. In collar thrust bearings for horizontal shafts the design is such that the bearing for each collar is separate and adjustable. The pressure on the different collars may thus be equalized.*

109. Lubrication of Journals. — The best method of lubrication is that in which the rubbing surfaces are constantly submerged in a bath of lubricating fluid. This method should be employed whenever possible, if the pressure and surface velocity are high. Unfortunately it cannot be used in the majority of cases. Let J, Fig. 121, represent a journal with its box, and let A, B, and C be oil holes. If oil be introduced into the hole A, it will tend to flow out from between the rubbing surfaces by the shortest way; *i. e.*, it will come out at D. A small amount will probably go toward the other end of the box because of capillary attraction, but usually none of it will reach the middle of the box. If oil be introduced at C, it will come out at E. A constant feed, therefore, might be maintained at A and C, and yet the middle of the box might run dry. If the oil be introduced at B, however, it tends to flow equally in both directions, and the entire journal is lubricated. The conclusion follows that oil ought, when possible, to be introduced at the middle of the length of a cylindrical journal. If a conical journal runs at a high velocity, the oil under the influence of centrifugal force tends to go to the large end of the cone, and therefore the oil should be introduced at the small end to insure its distribution over the entire journal surface.

If the end of a vertical thrust journal, whose outline is a cone or tractrix, as in Fig. 122, dips into a bath of oil, B, the oil will be

* For complete and varied details of marine thrust bearings see "Maw's Modern Practice in Marine Engineering."

carried by its centrifugal force, if the velocity be high, up between the rubbing surfaces, and will be delivered into the groove AA. If holes connect A and B, gravity will return the oil to B, and a constant circulation will be maintained. If the thrust journal has simply a flat end, as in Fig. 123, the oil should be supplied at the centre of the bearing; centrifugal force will then distribute it over the entire surface. Vertical shaft thrust journals may usually be arranged to run in an oil bath. Marine collar thrust journals are always arranged to run in an oil bath.

Sometimes a journal is stationary and the box rotates about it, as in the case of a loose pulley, Fig. 124. If the oil is introduced into a tube A, as is often done, its centrifugal force will carry it away from the rubbing surface. But if a hole is drilled in the axis of the journal, the lubricant introduced into it will be carried to the rubbing surfaces as required. If a journal is carried in a rotating part at a considerable distance from the axis of rotation, and it requires to be oiled while in motion, a channel may be provided from the axis of rotation where oil may be introduced conveniently, to the rubbing surfaces, and the oil will be carried out by centrifugal force. Thus Fig. 125 shows an engine crank in section. Oil is introduced at O, and centrifugal force carries it through the channel provided to a, where it serves to lubricate the rubbing surfaces of the crank-pin and its box. If a journal is carried in a reciprocating machine part, and requires to be oiled while in motion, the "wick and wiper" method is one of the best. See Fig. 126. An ordinary oil cup with an adjustable feed is mounted in a proper position opposite the end of the stroke of the reciprocating part, and a piece of flat wick projects from its delivery tube. A drop of oil runs down and hangs suspended at its end. Another oil cup is attached to the reciprocpting part, which carries a hooked "wiper," B. The delivery tube from C leads to the rubbing surfaces to be lubricated. When the reciprocating part reaches the end of its stroke the wiper picks off the drop of oil from the wick, and it runs down into the oil cup C, and thence to the surfaces to be lubricated. This method applies to the oiling of the cross-head pin

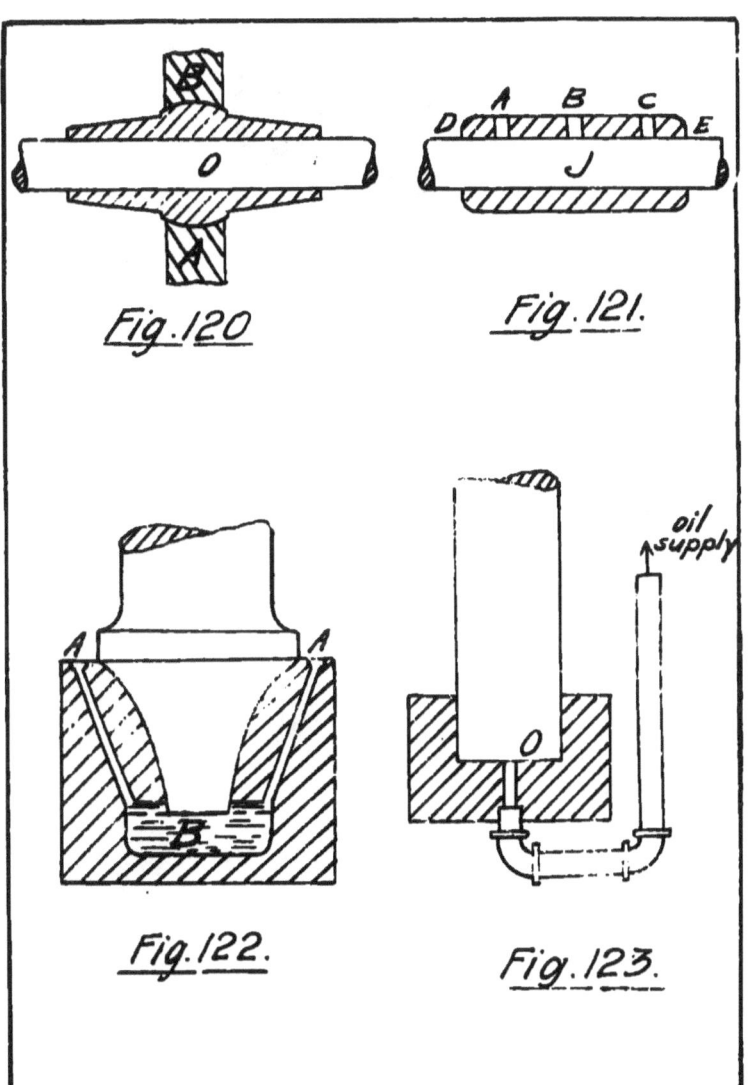

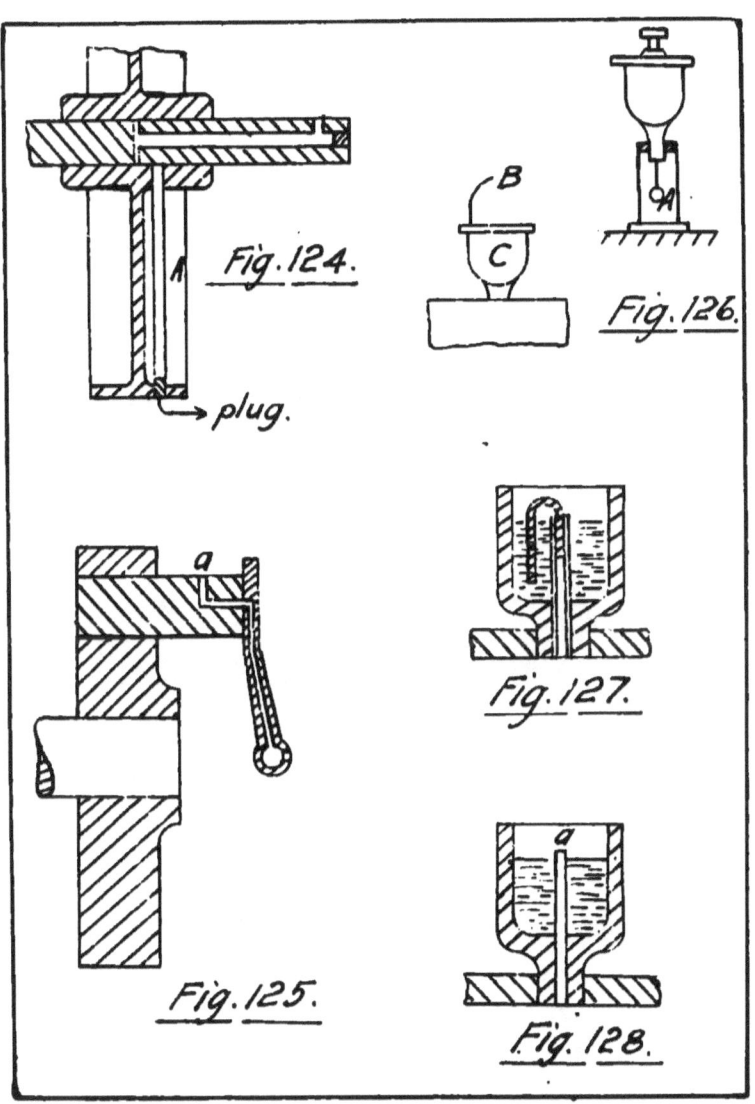

DESIGN OF JOURNALS. 125

of a steam engine. The same method is sometimes applied to the crank-pin, but here, through a part of the revolution, the tendency of the centrifugal force is to force the oil out of the cup, and therefore the plan of oiling from the axis is probably preferable.

When journals are lubricated by feed oilers, and are so located as not to attract attention if the lubrication should fail for any reason, "tallow boxes" are used. These are cup-like depressions usually cast in the box cap, and communicating by means of an oil hole with the rubbing surface. These cups are filled with grease that is solid at the ordinary temperature of the box, but if there is the least rise of temperature because of the failure of the oil supply, the grease melts and runs to the rubbing surfaces, and supplies the lubrication temporarily. This safety device is used very commonly on line shaft journals.

The most common forms of feed oilers are: I. The oil cup with an adjustable valve that controls the rate of flow. II. The oil cup with a wick feed, Fig. 127. The delivery has a tube inserted in it which projects nearly to the top of the cup. In this tube a piece of wicking is inserted, and its end dips into the oil in the cup. The wick, by capillary attraction, carries the oil slowly and continuously over through the tube to the rubbing surfaces. III. The cup with a copper rod, Fig. 128. The oil cup is filled with grease that melts with a very slight elevation of temperature, and A is a small copper rod dropped into the delivery tube and resting on the surface of the journal. The slight friction between the rod and the journal warms the rod and it melts the grease in contact with it, which runs down the rod to the rubbing surface. IV. Sometimes a part of the surface of the bottom half of the box is cut away and a felt pad is inserted, its bottom being in contact with an oil bath. This pad rubs against the surface of the journal, is kept constantly soaked with oil, and maintains lubrication.

CHAPTER XI.

SLIDING SURFACES.

110. So much of the accuracy of action of machines depends on the sliding surfaces that their design deserves the most careful attention. The perfection of the cross-sectional outline of the cylindrical or conical forms produced in the lathe, depends on the perfection of form of the spindle. But the perfection of the outlines of a section through the axis depends on the accuracy of the sliding surfaces. All of the surfaces produced by planers, and most of those produced by milling machines, are dependent for accuracy on the sliding surfaces in the machine.

Suppose that the short block A, Fig. 129, is the slider of a slider-crank chain, and that it slides on a relatively long guide, D. The direction of rotation of the crank, a, is as indicated by the arrow. B and C are the extreme positions of the slider. The pressure between the slider and the guide is greatest at the mid-position, A, and at the extreme positions, B and C, it is only the pressure due to the weight of the slider. Also the velocity is a maximum when the slider is in its mid-position, and decreases toward the ends, becoming zero when the crank a is on its centre. The work of friction is therefore greatest at the middle, and is very small near the ends. Therefore the wear would be greatest at the middle, and the guide would wear concave. If now the accuracy of a machine's working depends on the perfection of A's rectilinear motion, that accuracy will be destroyed as the guide D wears. Suppose a gib, EFG, to be attached to A, Fig. 130, and to engage with D, as shown, to prevent vertical looseness between A and D. If this gib be taken up to compensate wear after it has occurred, it will be loose in the middle position when it is tight at the ends, because of the unequal wear. Suppose that A and D are made of equal length, as in Fig. 131. Then when A is in the mid-position corresponding to maximum pressure, velocity, and wear, it is in contact with D throughout its entire surface, and the wear is therefore the same in all parts of the surface. The slider retains its accuracy of rectilinear motion regardless of the amount of wear, the gib may be set up, and will be equally tight in all positions.

If A and B, Fig. 132, are the extreme positions of a slider, D being the guide, a shoulder would be finally worn at C. It would be better to cut away the material of the guide, as shown by the dotted line. Slides should always "wipe over" the ends of the guide when it is possible. Sometimes it is necessary to vary the length of stroke of a slider, and also to change its position relatively to the guide. Examples: "Cutter bars" of slotting and shaping machines. In some of these positions, therefore, there will be a tendency to wear shoulders in the guide and also in the cutter bar itself. This difficulty is overcome if the slide and guide are made of equal length, and the design is such that when it is necessary to change the position of the cutter bar that is attached to the slide, the position of the guide may be also changed so that the relative position of slide and guide remains the same. The slider surface will then just completely cover the surface of the guide in the mid-position, and the slider will wipe over each end of the guide, whatever the length of the stroke.

In many cases it is impossible to make the slider and guide of equal length. Thus a lathe carriage cannot be as long as the bed; a planer table cannot be as long as the planer bed, nor a planer saddle as long as the cross-head. When these conditions exist especial care should be given to the following: I. The bearing surface should be made so large in proportion to the pressure to be sustained that the maintenance of lubrication shall be insured under all conditions. II. The parts which carry the wearing surfaces should be made so rigid that there shall be no possibility of the localization of pressure from yielding.

111. As to form, guides may be divided into two classes : angular guides and flat guides. Fig. 133 *a*, shows an angular guide, the pressure being applied as shown. The advantage of this form is, that as the rubbing surfaces wear, the slide follows down and takes up both the vertical and lateral wear. The objection to this form is that the pressure is not applied at right angles to the wearing surfaces, as it is in the flat guide shown in *b*. But in *b* a gib, *A*, must be provided to take up the lateral wear. The gib is either a wedge or a strip with parallel sides backed up by screws. Guides of these forms are used for planer tables. The weight of the table itself holds the surfaces in contact, and if the table is light the tendency of a heavy side cut would be to force the table up one of the angular surfaces away from the other. If the table is very heavy, however, there is little danger of this, and hence the angular guides of large planers are much flatter than those of smaller ones. In some cases one of the guides of a planer table is angular and the other is flat. The side bearings of the flat guide may then be omitted, as the lateral wear is taken up by the angular guide. This arrangement is undoubtedly good if both guides wear down equally fast.

112. Fig. 134 shows three forms of sliding surfaces such as are used for the cross slide of lathes, the vertical slide of shapers, the table slide of milling machines, etc. *A* is a taper gib that is forced in by a screw at *D* to take up wear. When it is necessary to take up wear at *B*, the screw may be loosened and a shim or liner may be inserted between the surfaces at *a*. *C* is a thin gib, and the wear is taken up by means of several screws like the one shown. This form is not so satisfactory as the wedge gib, as the bearing is chiefly under the points of the screws, the gib being thin and yielding, whereas in the wedge there is complete contact between the metallic surfaces.

113. The sliding surfaces thus far considered have to be designed so that there will be no lost motion while they are moving, because they are required to move while the machine is in operation. The gibs have to be carefully designed and accurately set so that the

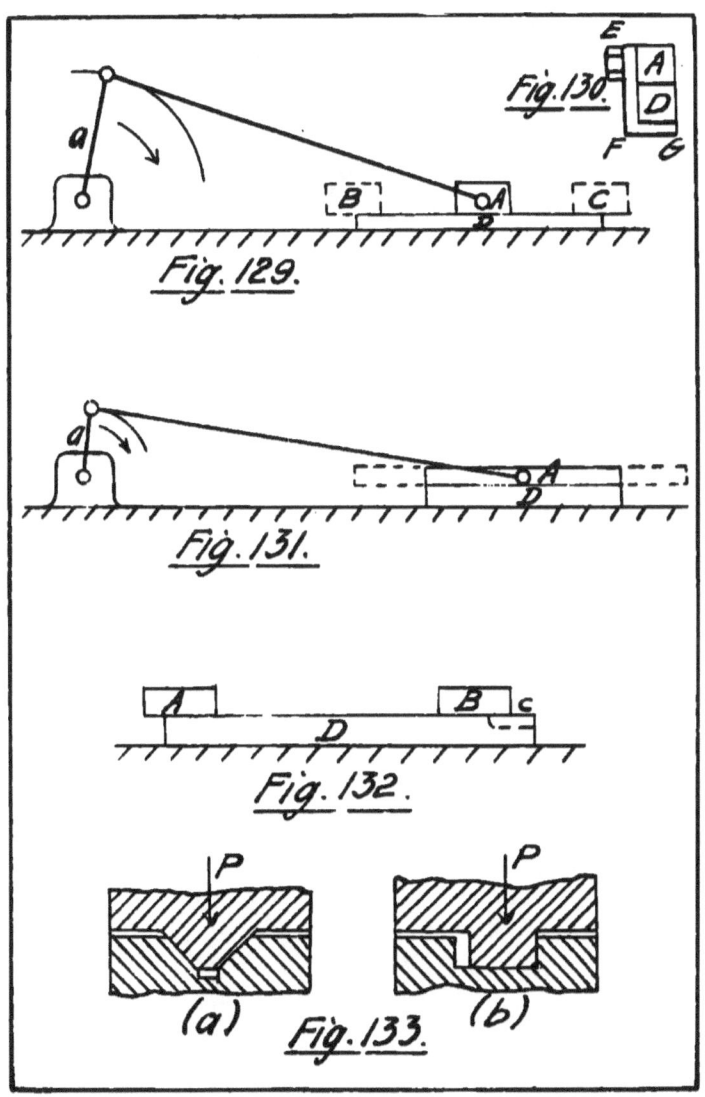

moving part shall be just "tight and loose," *i. e.*, so that it shall be free to move, without lost motion to interfere with the accurate action of the machine. There is, however, another class of sliding parts, like the sliding head of a drill press, or the tail stock of a lathe, that are never required to move while the machine is in operation. It is only required that they shall be capable of being fastened accurately in a required position, their movement being simply to readjust them to other conditions of work, while the machine is at rest. No gib is necessary and no accuracy of *motion* is required. It is simply necessary to insure that their position is accurate when they are clamped for the special work to be done.

CHAPTER XII.

BOLTS AND SCREWS AS MACHINE FASTENINGS.

114. *Classification* may be made as follows: I. Bolts. II. Studs. III. Cap Screws, or Tap Bolts. IV. Set Screws. V. Machine Screws.

A "bolt" consists of a head and round body on which a thread is cut, and upon which a nut is screwed. When a bolt is used to connect machine parts, a hole the size of the body of the bolt is drilled entirely through both parts, the bolt is put through, and the nut screwed down upon the washer. See Fig. 135.

A "stud" is a piece of round metal with a thread cut upon each end. One end is screwed into a tapped hole in some part of a machine, and the piece to be held against it, having a hole the size of the body of the stud, is put on, and a nut is screwed upon the other end of the stud against the piece to be held. See Fig. 136.

A "cap screw" is a substitute for a stud, and consists of a head and body on which a thread is cut. See Fig. 137. The screw is passed through the removable part and screwed into a tapped hole in the part to which it is attached. A cap screw is a stud with a head substituted for the nut.

A hole should never be tapped into a cast iron machine part when it can be avoided. Cast iron is not good material for the thread of a nut, since it is weak and brittle and tends to crumble. In very many cases, however, it is absolutely necessary to tap into cast iron. It is then better to use studs, if the attached part needs to be removed often, because studs are put in once for all, and the cast iron thread would be worn out eventually if cap screws were used.

When one machine part surrounds another, as a pulley hub

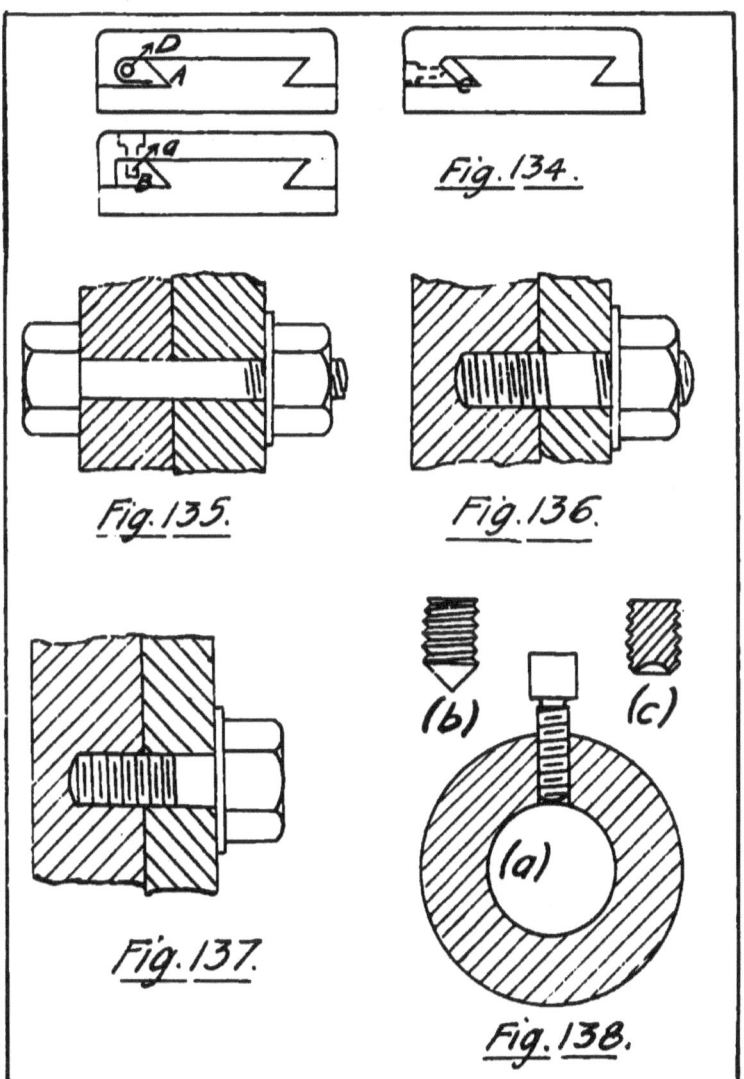

surrounds a shaft, relative motion of the two is often prevented by means of a "set screw" which is a threaded body with a small square head. Fig. 138. The end is either pointed as in Fig. 138 b, or cupped as in c, and is forced against the inner part by screwing through a tapped hole in the outer part.

The term "machine screws" covers many forms of small screws, usually with screw-driver heads. All of the kinds given in this classification are made in great variety of size, form, length, etc.

115. Design of Bolts and Screws.—Tensile stress is induced in a bolt by tightening it. This stress may equal, or very greatly exceed, the tensile stress due to working forces. The stress due to tightening may be approximately found as follows: In Fig. 139, a force P is applied to the wrench handle at A. During the turning of the wrench the ratio of movement of the point A to the movement of the nut axially $= 2\pi l$ to p, in which l is the wrench lever arm, and p is the pitch of the screw (axial distance between threads). If there were no frictional resistance, the force P and the resistance to tension of the bolt, $= T$, would be in equilibrium at the instant of tightening up, and the following equation would be true:

$$P2\pi l = Tp.$$

But force P must also overcome the frictional resistance between the nut and the screw thread, and between the nut and washer. This resistance $R = Tf$; in which $f =$ the coefficient of friction. The radius, r, of this resistance, R, may be assumed, for this approximation, equal to the radius of the top of the screw threads $\times$ 1·5. The ratio of movement of R to axial movement of the nut $= 2\pi r$ to p. At the instant of tightening up there is equilibrium between P, R, and T, and the following equation is true:

$$2\pi p l = Tp + 2\pi Rr \text{ ; substituting for } R = fT,$$
$$= Tp + 2\pi Tfr,$$

whence
$$T = \frac{2\pi Pl}{p + 2\pi fr}.$$

For a ½" bolt the values are $l = 8"$, $p = 0.077"$, $f = 0.15$, $r = 0.375$. Making $P = 1$ lb., $T = 116$ lbs. Hence, for every pound applied at A there results 116 lbs. tensile stress in the bolt. The ultimate strength of the bolt ÷ 116 equals the force applied at A necessary to break the bolt. The area of cross-section at the bottom of the thread of a ½" bolt $= 0.12$ sq. in. Assume the ultimate strength of the material of the bolt $= 50000$ lbs. per sq. in. The ultimate tensile strength of the bolt $= 50000 \times 0.12 = 6000$ lbs. Then the force at A to break the bolt $= 6000 \div 116 = 52$ lbs. nearly. This is probably not the actual force that would break the bolt, because the assumptions are probably somewhat inaccurate; but it indicates that a half inch bolt may be pulled in two by force applied by a man to the wrench handle. For a ⅜" bolt the force becomes about 100 lbs.

Suppose a nut screwed up with a resulting tensile stress in the bolt $= p$. Suppose that a gradually increasing working force, $= p_1$, is applied. If there were no elongation of the bolt, the total stress in the bolt would equal $p + p_1$. But elongation does result from the application of p_1, and p is reduced, and the total stress in the bolt is less than $p + p_1$.

116. Illustration.—In Fig. 140 the tensile stress in the bolt due to screwing up $= p$. The pressure between the surfaces in contact at CD is therefore $= p$. Suppose a working force, p_1, applied tending to separate A and B. The bolt yields to the increased stress, and the pressure at CD is reduced. The tensile stress in the bolt is now equal to the working force plus the reduced pressure at CD. When the working force reduces the pressure at CD to zero, the stress in the bolt $=$ the working force, and if CD were a steam joint, it would leak.

117. It is required to design the fastenings to hold on the steam chest cover of a steam engine. The opening to be covered is rectangular, 10"x 12". The maximum steam pressure is 100 lbs. per square inch. The joint must be held steam tight. Short unyielding fastenings are therefore best suited to the purpose, and studs will be used. They will be made of machinery steel of 60000 lbs.

BOLTS AND SCREWS AS MACHINE FASTENINGS. 133

tensile strength, and will be $\frac{3}{4}"$ outside diameter because smaller studs may be ruptured by the force applied in screwing up. It will be assumed that the stress on the studs is equal to $p + p_1$, i. e., the stress due to screwing up, plus the working stress. This assumption cannot be exactly true, as is seen from the preceding illustration; but the resulting error is on the safe side. The diameter of a $\frac{3}{4}"$ stud at the bottom of the thread is $0.62"$, and the area is $= 0.62^2 \times \pi \div 4 = 0.3$ sq. in. The ultimate strength of the stud is, therefore, $0.3 \times 60000 = 18000$. The factor of safety may be 4 because the stress member is of resilient material and is not subject to shocks. Then the allowable stress on each stud would be equal to $18000 \div 4 = 4500$. The stress due to screwing up, subtracted from the total allowable stress gives the allowable working stress in the stud. The stress due to screwing up may be found from the above equation for T. Assuming $P = 30$ lbs.; $l = 10"$: $f = 0.15$; $r = 0.56$; gives $T = 3000$ lbs. nearly. The allowable working stress in each stud equals $4500 - 3000 = 1500$ lbs. The maximum working force on the cover equals the area of the opening covered, multiplied by the maximum working pressure per square inch; $= 10 \times 12 = 120$ sq. in. $\times 100$ lbs. per sq. in. $= 12000$ lbs. This divided by the allowable working pressure for each stud gives the number of studs required for strength, $= 12000 \div 1500 = 8$. Therefore 8 studs will serve for strength. But in order to make a steam tight joint, with a reasonable thickness of steam chest cover, the distance between the stud centres should not be greater than about $4\frac{1}{2}"$. The opening is $10" \times 12"$, as shown in Fig. 141. There must be a band about $\frac{3}{8}"$ wide around this for making the joint upon which the studs must not encroach. This makes the distance between the vertical rows of studs $14"$, and between the horizontal rows $12"$. The whole length over which the studs are to be distributed then $= 12 + 12 + 14 + 14 = 52"$. If they are $4.5"$ apart the number of studs $= 52 \div 4.5 = 11.5$. Hence it is necessary to use 12 studs to make the joint tight, while 8 would serve for strength. In order to get a symmetrical arrangement, it will probably be necessary to use 14 studs. The number of studs is therefore dictated by the conditions necessary to maintain a tight steam joint, and not by the applied forces.

118. The elongation of a bolt with a given total stress, depends upon the *length* and *area* of its least cross-section. Suppose, to illustrate, that the bolt, Fig. 142, has a reduced section over a length l as shown. This portion A has less cross-sectional area than the rest of the bolt, and when any tensile force is applied, the resulting *unit* stress will be greater in A than elsewhere.. The unit strain, or elongation, will be proportionately greater, up to the elastic limit; and if the elastic limit is exceeded in the portion A, the elongation there will be far greater than elsewhere. If there is much difference of area and the bolt is tested to rupture, the elongation will be chiefly at A. There would be a certain elongation *per inch* of A at rupture. Hence, the greater the length of A, the greater the total elongation of the bolt. If the bolt had not been reduced at A, the minimum section would be at the root of the screw threads. The axial length of this section is very small. Hence the elongation at rupture would be small. Suppose there are two bolts, A with, and B without, the reduced section. They are alike in other respects. They are subjected to equal tensile shocks. Let the energy of the shock $= E$. This energy is divided into force and space factors by the resistance of the bolts. The space factor equals the elongation of the bolt. This is greater in A than in B because of the yielding of the reduced section. But the product of force and space factors is the same in both bolts, $= E$; hence the resulting stress in the minimum section is less for A than for B. The stress in A may be less than the breaking stress; while the greater stress in B may break it. *The capacity of the bolt to resist shock is therefore increased by lengthening its minimum section to increase the yielding and reduce stress.* This is not only true of bolts, but of all stress members in machines.

The whole body of the bolt might have been reduced, as shown by the dotted lines in Fig. 142, with resulting increase of capacity to resist shock. Turning down a bolt, however, weakens it to

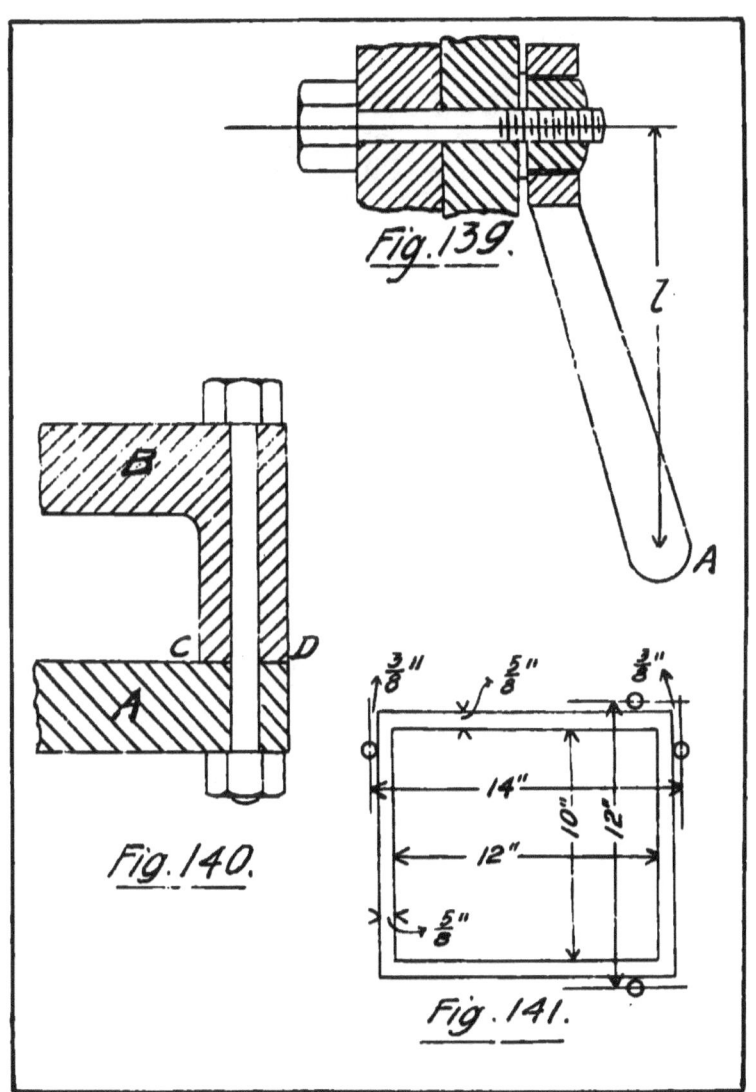

Fig. 139.

Fig. 140.

Fig. 141.

resist torsion and flexure, because it takes off the material which is most effective in producing large polar and rectangular moments of inertia of cross-section. If the cross-sectional area is reduced by drilling a hole as shown in Fig. 143, the torsional and transverse strength is but slightly decreased, but the elongation will be as great, with the same area, as if the area had been reduced by turning down.

119. Prof. Sweet had a set of bolts prepared for special test. The bolts were $1\frac{1}{4}''$ diameter and about $12''$ long. They were made of high grade wrought iron, and were duplicates of the bolts used at the crank end of the connecting-rods of one of the standard sizes of the Straight Line Engine. Half of the bolts were left solid, while the other half were carefully drilled to give them uniform cross-sectional area throughout. The tests were made under the direction of Prof. Carpenter at the Sibley College Laboratory. One pair of bolts was tested to rupture by tensile force gradually applied. The undrilled bolt broke in the thread with a total elongation of $0.25''$. The drilled bolt broke between the thread and the bolt head with a total elongation of $2.25''$. If it be assumed that the mean force applied was the same in both cases, it follows that the total resilience of the drilled bolt was nine times as great as that of the solid one. " Drop tests," *i. e.*, tests which brought tensile shock to bear upon the bolts, were made on other similar pairs of bolts, which tended to confirm the general conclusion.

120. It is required to design proper fastenings for holding on the cap of a connecting-rod like that shown in Fig. 144. These fastenings are required to sustain shocks, and may be subjected to a maximum accidental stress of 20000 lbs. There are two fastenings, and therefore each must be capable of sustaining safely a stress of 10000 lbs. They should be designed to yield as much as is consistent with strength ; in other words, they should be tensile springs to cushion shocks and thereby reduce the resulting force they have to sustain. Bolts should therefore be used, and the weakest section should be made as long as possible. Wrought iron will be used whose tensile strength is 50000 lbs. per square inch.

The stress given is the maximum accidental stress, and is four times the working stress. It is not, therefore, necessary to give the bolts great excess of strength over that necessary to resist actual rupture by the accidental force. Let the factor of safety be 2. Then the cross-sectional area of each bolt must be such that it will just sustain $10000 \times 2 = 20000$ lbs. This area $= 20000 \div 50000 = 0.4$ square inches. This area corresponds to a diameter of 0.71", and that is nearly the diameter of a $\frac{7}{8}$" bolt at the bottom of the thread; hence $\frac{7}{8}$" bolts will be used. The cross-sectional area of the body of the bolt must now be made at least as small as that at the bottom of the thread. This may be accomplished by drilling.

121. When bolts are subjected to constant vibration there is a tendency for the nuts to loosen. There are many ways to prevent this, but the most common one is by the use of jamb nuts. Two nuts are screwed on the bolt; the under one is set up against the surface of the part to be held in place, and then while this nut is held with a wrench the other nut is screwed up against it tightly. Suppose that the bolt has its axis vertical and that the nuts are screwed on the upper end. The nuts being screwed against each other the upper one has its internal screw surfaces forced against the under screw surfaces of the bolt, and if there is any lost motion, as there almost always is, there will be no contact between the upper surfaces of the screw on the bolt and the threads of the nut. Just the reverse is true of the under nut; *i. e.*, there is no contact between the under surfaces of the threads on the bolt and the threads on the nut. Therefore no pressure that comes from the under side of the under nut can be communicated to the bolt through the under nut directly, but it must be received by the upper nut and communicated by it to the bolt, since it is the upper nut alone that has contact with the under surfaces of the thread. Therefore the jamb nut, which is usually made about half as thick as the other, should always be put on next to the surface of the piece to be held in place.

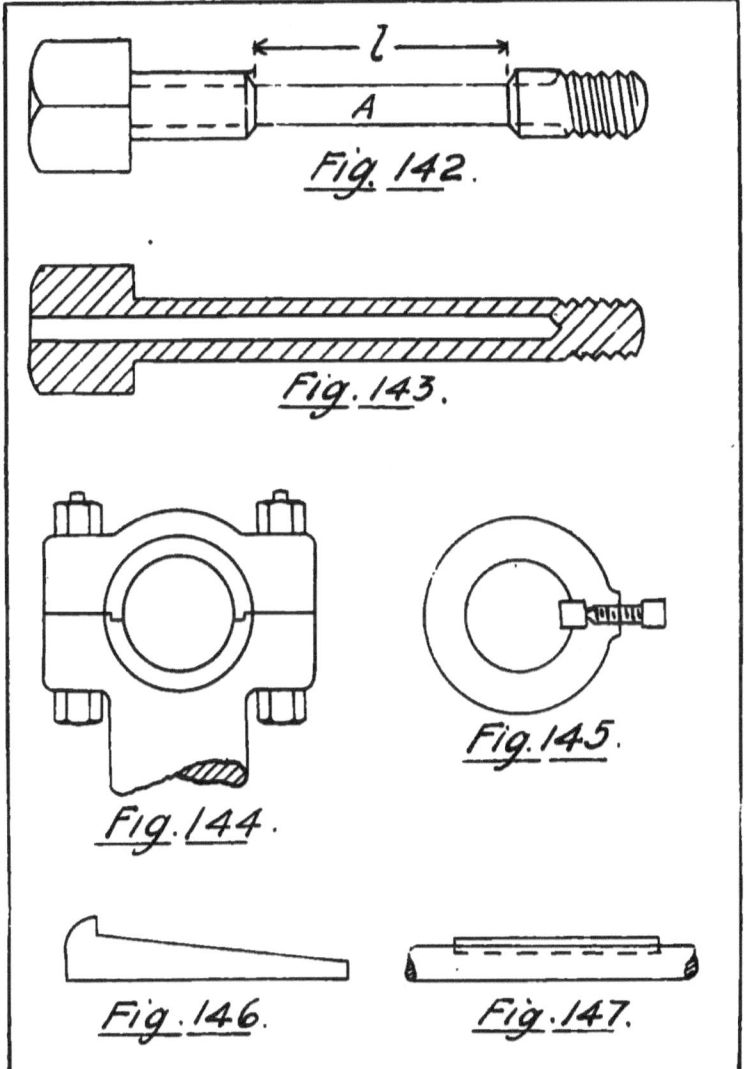

Fig. 142.

Fig. 143.

Fig. 144.

Fig. 145.

Fig. 146.

Fig. 147.

CHAPTER XIII.

MEANS FOR PREVENTING RELATIVE ROTATION.

122. Keys are chiefly used to prevent relative rotation between shafts and the pulleys, gears, etc. which they support. Keys may be divided into parallel keys, taper keys, and feathers or splines.

For a **parallel key** the "seat," both in the shaft and the attached part, has parallel sides, and the key simply prevents relative rotary motion. Motion parallel to the axis of the shaft must be prevented by some other means; as by set screws which bear upon the top surface of the key, as shown in Fig. 145. A parallel key should fit accurately on the sides and loosely at the top and bottom.

A **taper key** has parallel sides and has its top and bottom surfaces tapered, and is made to fit on all four surfaces, being driven tightly "home." It prevents relative motion of any kind between the parts connected. If a key of this kind has a head, as shown in Fig. 146, it is called a "draw key," because it is drawn out when necessary, by driving a wedge between the hub of the attached part and the head of the key. When a taper key has no head it is removed by driving against the point with a "key drift."

Feathers or **Splines** are keys that prevent relative rotation, but purposely allow axial motion. They are sometimes made fast in the shaft, as in Fig. 147, and there is a key "way" in the attached part that slides along the shaft. Sometimes the feather is fastened in the hub of the attached part, as shown in Fig. 148, and slides in a long key way in the shaft.

John Richards' rule for keys is (see Fig. 149) $w = \dfrac{d}{4}$. t has such value that $a = 30°$. This rule is deviated from somewhat, as shown by the following table taken from Richards' "Manual of Machine Construction," page 58.

$w=$	1	$1\tfrac{1}{4}$	$1\tfrac{1}{2}$	$1\tfrac{3}{4}$	2	$2\tfrac{1}{2}$	3	$3\tfrac{1}{2}$	4	5	6	7	8
$d=$	$\tfrac{1}{4}$	$\tfrac{5}{16}$	$\tfrac{3}{8}$	$\tfrac{7}{16}$	$\tfrac{1}{2}$	$\tfrac{5}{8}$	$\tfrac{3}{4}$	$\tfrac{7}{8}$	1	$1\tfrac{1}{8}$	$1\tfrac{3}{8}$	$1\tfrac{1}{2}$	$1\tfrac{3}{4}$
$t=$	$\tfrac{5}{32}$	$\tfrac{3}{16}$	$\tfrac{1}{4}$	$\tfrac{9}{32}$	$\tfrac{5}{16}$	$\tfrac{3}{8}$	$\tfrac{7}{16}$	$\tfrac{1}{2}$	$\tfrac{5}{8}$	$\tfrac{11}{16}$	$\tfrac{13}{16}$	$\tfrac{7}{8}$	1

When d exceeds 8" two or more keys should be used, and w may then $= d \div 16$; t being as before of such value that a shall $= 30°$. The following table for dimensions for parallel keys is also from Richards' "Manual":

$d=$	1	$1\tfrac{1}{4}$	$1\tfrac{1}{2}$	$1\tfrac{3}{4}$	2	$2\tfrac{1}{2}$	3	$3\tfrac{1}{2}$	4
$w=$	$\tfrac{5}{32}$	$\tfrac{7}{32}$	$\tfrac{9}{32}$	$\tfrac{11}{32}$	$\tfrac{13}{32}$	$\tfrac{15}{32}$	$\tfrac{17}{32}$	$\tfrac{9}{16}$	$\tfrac{11}{16}$
$t=$	$\tfrac{3}{16}$	$\tfrac{1}{4}$	$\tfrac{5}{16}$	$\tfrac{3}{8}$	$\tfrac{7}{16}$	$\tfrac{1}{2}$	$\tfrac{9}{16}$	$\tfrac{5}{8}$	$\tfrac{3}{4}$

Also this for feathers:

$d=$	$1\tfrac{1}{4}$	$1\tfrac{1}{2}$	$1\tfrac{3}{4}$	2	$2\tfrac{1}{4}$	$2\tfrac{1}{2}$	3	$3\tfrac{1}{2}$	4	$4\tfrac{1}{2}$
$w=$	$\tfrac{1}{4}$	$\tfrac{1}{4}$	$\tfrac{5}{16}$	$\tfrac{5}{16}$	$\tfrac{3}{8}$	$\tfrac{3}{8}$	$\tfrac{1}{2}$	$\tfrac{9}{16}$	$\tfrac{9}{16}$	$\tfrac{5}{8}$
$t=$	$\tfrac{3}{8}$	$\tfrac{3}{8}$	$\tfrac{7}{16}$	$\tfrac{7}{16}$	$\tfrac{1}{2}$	$\tfrac{1}{2}$	$\tfrac{5}{8}$	$\tfrac{3}{4}$	$\tfrac{3}{4}$	$\tfrac{7}{8}$

For keying hand wheels and other parts that are not subjected to very great stress, a cheap and satisfactory method is to use a round key driven into a hole drilled in the joint, as in Fig. 150. If the two parts are of different material, one much harder than the other, this method should not be used, as it is almost impossible in such case to make the drill follow the joint.

The taper of keys varies from $\tfrac{1}{8}$" to $\tfrac{1}{4}$" to the foot.

A *cotter* is a key that is used to attach parts subjected to a force of tension tending to separate them. Thus piston rods are often connected to both piston and cross-head in this way. Also the sections of long pump-rods, etc.

Fig. 151 shows machine parts held against tension by cotters. It is seen that the joint may yield by shearing the cotter at AB

and CD; or by shearing CPQ and ARS; by shearing on the surfaces MO and LN; or by tensile rupture of the rod on a horizontal section at LM. All of these sections should be sufficiently large to resist the maximum stress safely. The difficulty is usually to get LM strong enough in tension: but this may usually be accomplished by making the rod larger, or the cotter thinner and wider. It is found that taper surfaces if they be smooth and somewhat oily will just cease to stick together when the taper equals 1·5″ per foot. The taper of the rod in Fig. 151 should be about this value in order that it may be removed conveniently when necessary.

123. Shrink and Force Fits.— Relative rotation between machine parts is also prevented sometimes by means of shrink and force fits. In the former the shaft is made larger than the hole in the part to be held upon it, and the metal surrounding the hole is heated, usually to low redness, and because of the expansion it may be put on the shaft and on cooling, it shrinks and "grips" the shaft. A key is sometimes used in addition to this.

Force fits are made in the same way except that they are put together cold, either by driving together with a heavy sledge or by forcing together by hydraulic pressure. The necessary allowance for forcing, *i. e.*, the excess of shaft diameter over the diameter of the hole, is given in the following table:

	Inches.									
Diameter of Shaft	1	2	3	4	5	6	7	8	9	10
Allowance for Forcing	0·004	0·005	0·006	0·006	0·007	0·008	0·008	0·009	0·01	0·01

Experience shows that with this allowance a steel shaft may be forced into a hole in cast iron by a total pressure of from 40 to 90 tons. There is no need of keying when parts are put together in this way.

CHAPTER XIV.

FORM OF MACHINE PARTS AS DICTATED BY STRESS.

124. Suppose that A and B, Fig. 152, are two surfaces in a machine to be joined by a member subjected to simple tension. What is the proper form for the member? The stress in all sections of the member at right angles to the line of application, AB, of the force, will be the same. Therefore the areas of all such sections should be equal; hence the outlines of the member should be straight lines parallel to AB. The distance of the material from the axis AB has no effect on its ability to resist tension. Therefore there is nothing in the character of the stress that indicates the form of the cross-section of the member. The form most cheaply produced, both in the rolling mill and the machine shop, is the cylindrical form. Economy, therefore, dictates the circular cross-section. After the required area necessary for safely resisting the stress is determined, it is only necessary to find the corresponding diameter, and it will be the diameter of all sections of the required member if they are made circular. Sometimes in order to get a more harmonious design, it is necessary to make the tension member just considered of rectangular cross-section, and this is allowable although it almost always costs more. The thin, wide, rectangular section should be avoided, however, because of the difficulty of insuring a uniform distribution of stress. A unit stress might result from this at one edge, greater than the strength of the material, and the piece would yield by tearing, although the *average* stress might not have exceeded a safe value.

If the stress be compression instead of tension, the same considerations dictate its form as long as it is a "short block," *i. e.*, as long

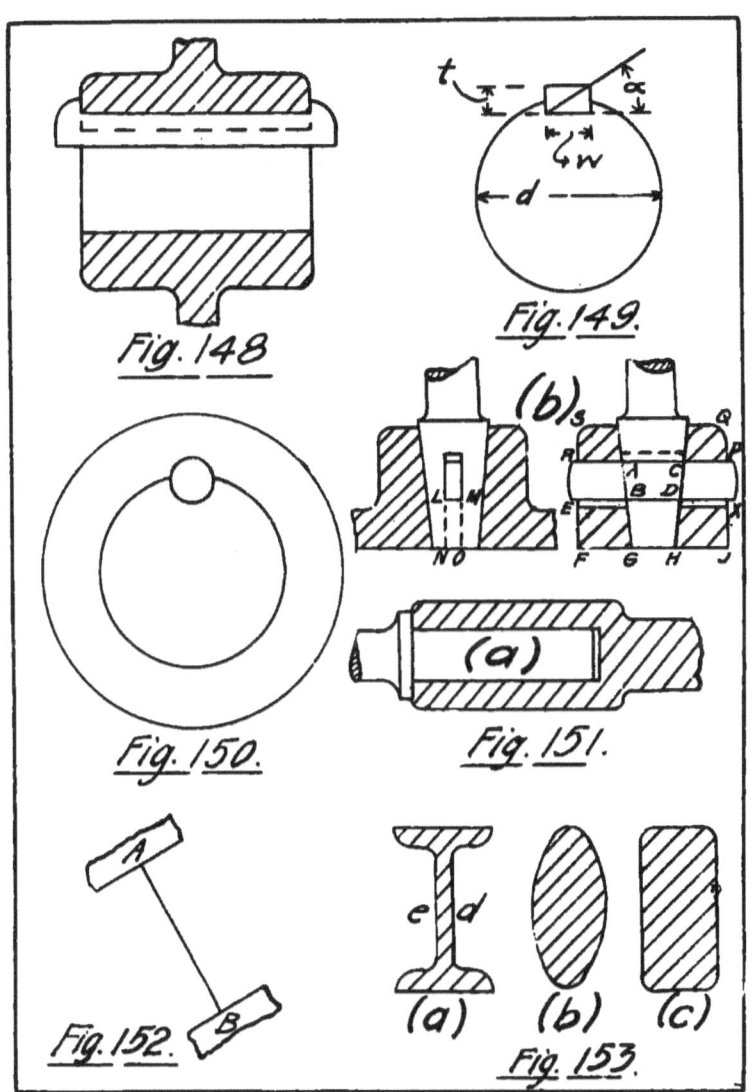

Fig. 148

Fig. 149.

Fig. 150.

Fig. 151.

Fig. 152.

Fig. 153.

FORM OF MACHINE PARTS AS DICTATED BY STRESS. 141

as the ratio of length to lateral dimensions is such that it is sure to yield by crushing instead of by "buckling." A short block, therefore, should have its longitudinal outlines parallel to its axis, and its cross-section may be of any form that economy or appearance may dictate. Care should be taken, however, that the *least* lateral dimension of the member be not made so small that it is thereby converted into a "long column."

If the ratio of longitudinal to lateral dimensions is such that the member becomes a "long column," the conditions that dictate the form are changed, because it would yield by buckling or flexure, instead of crushing. The strength and stiffness of a long column are proportional to the moment of inertia of the cross-section about a gravity axis at right angles to the plane in which the flexure occurs. A long column with "fixed" or "rounded" ends has a tendency to yield by buckling which is equal in all directions. Therefore the moment of inertia needs to be the same about all gravity axes, and this of course points to a circular section. Also the moment of inertia should be as large as possible for a given weight of material, and this points to the hollow section. The disposition of the metal in a circular hollow section is the most economical one for long column machine members with fixed or rounded ends. This form, like that for tension, may be changed to the rectangular hollow section if appearance requires such change. If the long column machine member be "pin connected," the tendency to buckle is greatest in a plane through the line of direction of the compressive force, and at right angles to the axis of the pins. The moment of inertia of the cross-section should therefore be greatest about a gravity axis parallel to the axis of the pins. Example: a steam engine connecting-rod.

When the machine member is subjected to transverse stress the best form of cross-section is probably the I section, *a*, Fig. 153, in which a relatively large moment of inertia, with economy of material, is obtained by putting the excess of the material where it is most effective to resist flexure, *i. e.*, at the greatest distance from the given gravity axis. Sometimes, however, if the I section has to be produced by cutting away the material at e and d, in the machine shop, instead of producing the form directly in the rolls, it is cheaper to use the solid rectangular section c. If the member subjected to transverse stress is for any reason made of cast material, as is often the case, the form b is preferable, for the following reasons: I. The best material is almost sure to be in the thinnest part of a casting, and therefore in this case is at f and g, where it is most effective to resist flexure. II. The pattern for the form b is more cheaply produced and maintained than that for a. III. If the surface is left without finishing from the mould, any imperfections due to the foundry work are more easily corrected in b than in a. Machine members subjected to transverse stress, which continually change their position relatively to the force that produces the flexure, should have the same moment of inertia about all gravity axes. As, for instance, rotating shafts that are strained transversely by the force due to the weight of a fly-wheel, or that due to the tension of a driving belt. The best form of cross-section in this case is circular. The hollow section would give the greatest economy of material, but hollow members are expensive to produce in wrought material, such as is almost invariably used for shafts. Hence the solid circular section is used.

125. Torsional strength and stiffness are proportional to the polar moment of inertia of the cross-section of the member. This is equal to the sum of the moments of inertia about two gravity axes at right angles to each other. The forms in Fig. 153 are therefore not correct forms for the resistance of torsion. The circular solid or hollow section, or the rectangular solid or hollow section, should be used.

The I section, Fig. 154, is a correct form for resisting the stress P, applied as shown. Suppose the web c to be divided on the line CD, and the parts to be moved out so that they occupy the positions shown at a and b. The form thus obtained is called a "box section." By making this change the moment of inertia about AB has not been changed, and therefore the new form is just as effective to resist flexure due to the force P as it was before the change. The box section is better able to resist torsional stress, because the change made to convert the I section into the box section has increased the polar moment of inertia. The two forms are equally good to resist tensile and compressive force if they are sections of short blocks. But if they are both sections of long columns, the box section would be preferable, because the moments of inertia would be more nearly the same about all gravity axes.

126. The framing of machines almost always sustains combined stresses, and if the combination of stresses include torsion, flexure in different planes, or long column compression, the box section is the best form. In fact the box section is by far the best form for the resisting of stress in machine frames. There are other reasons, too, beside the resisting of stress that favor its use.* I. Its appearance is far finer, giving an idea of completeness that is always wanting in the ribbed frames. II. The faces of a box frame are always available for the attachment of auxiliary parts without interfering with the perfection of the design. III. The strength can always be increased by decreasing the size of the core, without changing the external appearance of the frame, and therefore without any work whatever on the pattern itself. The cost of patterns for the two forms is probably not very different; the pattern itself being the more expensive in the ribbed form, and the necessary core boxes adding to the expense in the case of the box form. The expense of production in the foundry, however, is greater for the box form than for the ribbed form, because core work is more expensive than "green sand" work. The balance of advantage is very greatly in favor of box forms, and this is now being recognized in the practice of the best designers of machinery.

127. To illustrate the application of the box form to machine members, let the table of a planer be considered. The cross-section is almost universally of the form shown in Fig. 155. This is evidently a form that would yield easily to a force tending to twist it, or to a force acting in a vertical plane tending to bend it. Such forces may be brought upon it by "strapping down work," or by the

* See Richards' "Manual of Machine Construction."

support of heavy pieces upon centres. Thus in Fig. 156 the heavy piece E is supported between the centres. For proper support the centres need to be screwed in with a considerable force. This causes a reaction tending to separate the centres and to bend the table between C and D. As a result of this, the Vs on the table no longer have a bearing throughout the entire surface of the guides on the bed, but only touch near the ends, the pressure is concentrated upon small surfaces, the lubricant is squeezed out, the Vs and guides are "cut," and the planer is rendered incapable of doing accurate work. If the table were made of the box form shown in Fig. 157, with partitions at intervals throughout its length, it would be far more capable of maintaining its accuracy of form under all kinds of stress, and would be more satisfactory for the purpose for which it is designed.*

The bed of a planer is usually of the form shown in section in Fig. 158, the side members being connected by "cross girts" at intervals. This is evidently not the best form to resist flexure and torsion, and a planer bed may sustain both, either by reason of improper support, or because of changes in the form of foundation. If the bed were of box section with cross partitions, it would sustain greater stress without undue yielding. Holes could be left in the top and bottom to admit of supporting the core in the mould; to serve for the removal of the core sand; and to render accessible the gearing and other mechanism inside of the bed.

This same reasoning applies to lathe beds. They are strained transversely by force tending to separate the centres, as in the case of "chucking"; torsionally by the reaction of a tool cutting the surface of a piece of large diameter; and both torsion and flexure may result, as in the case of the planer bed, from an improperly designed or yielding foundation. The box form would be the best possible form for a lathe bed; some difficulties in adaptation, however, have prevented its extended use as yet.

These examples illustrate principles that are of very broad application in the designing of machines.

* Prof. Sweet has designed and constructed such a table for a large milling machine.

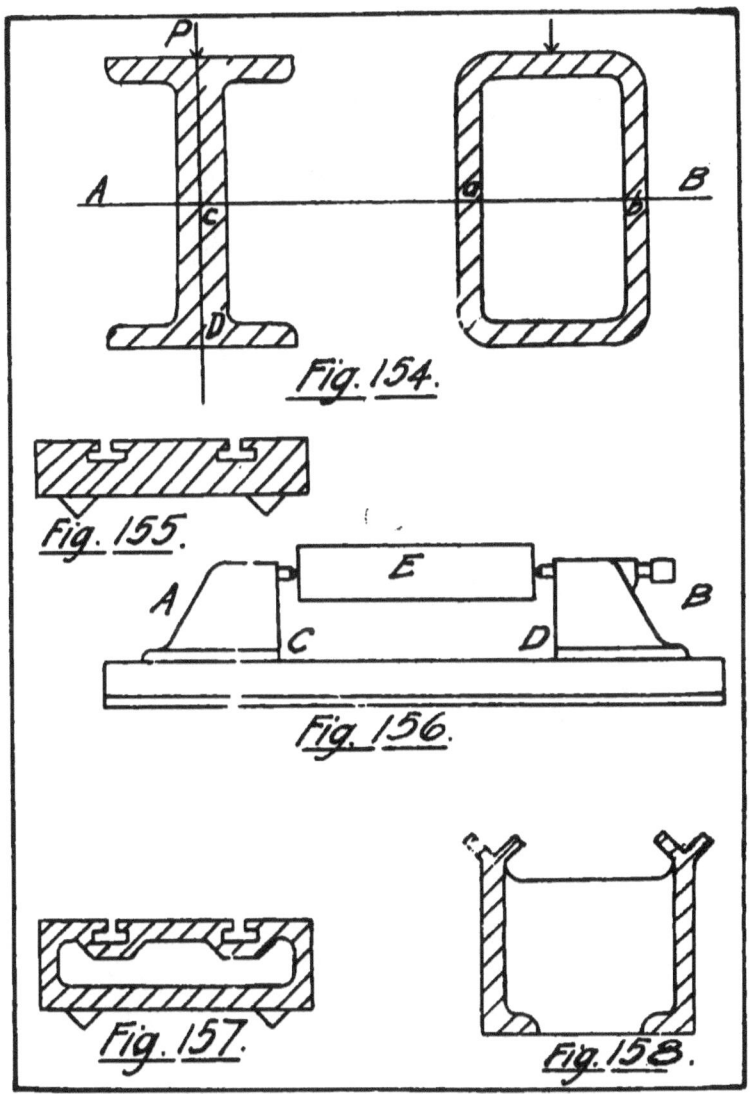

Fig. 154.

Fig. 155.

Fig. 156.

Fig. 157.

Fig. 158.

FORM OF MACHINE PARTS AS DICTATED BY STRESS. 145

128. Often in machines there is a part that projects either vertically or horizontally and sustains a transverse stress; it is a cantilever, in fact. If only transverse stress is sustained, and the thickness is uniform, the outline for economy of material is parabolic. In such a case, however, the outline curve of the member should start from the point of application of the force, and not from the extreme end of the member, as in the latter case there would be an excess of material. Thus in A, Fig. 159, P is the extreme position at which the force can be applied. The parabolic curve a is drawn from the point of application of P. The end of the member is supported by the auxiliary curve c. The curve b drawn from the end gives an excess of material. The curves a and c may be replaced by a single continuous curve as in C, or a tangent may be drawn to a at its middle point as in B, and this straight line used for the outline; the excess of material being slight in both cases. Most of the machine members of this kind, however, are subjected also to other stresses. Thus the "housings" of planers have to resist torsion and side flexure. They are very often supported by two members of parabolic outline; and, to insure the resistance of the torsion and side flexure, these two members are connected at their parabolic edges by a web of metal that really converts it into a box form. Machine members of this kind may also be supported by a brace, as in D. The brace is a compression member and may be stiffened against buckling by a "web" as shown, or by an auxiliary brace.

CHAPTER XV.

MACHINE SUPPORTS.

129. The Single Box Pillar Support is best and simplest for machines whose size and form admit of its use. When a support is a single continuous member, its design should be governed by the following principles:

I. The amount of material in the cross-section is determined by the intensity of the load. If vibrations are also to be sustained, the amount of material must be increased for this purpose.

II. The vertical centre line of the support should coincide with the vertical line through the centre of gravity of the part supported.

III. The vertical outlines of the support should taper slightly and uniformly on all sides. If they were parallel they would appear nearer together at the bottom.

IV. The external dimensions of the support must be such that the machine has the appearance of being in stable equilibrium. The outline of all heavy members of the machine supported must be either carried without break to the foundation, or if they overhang, must be joined to the support by means of parabolic outlines, or by the straight lines of the brace form.

Illustration. — In Fig. 160 the first three principles may be fulfilled, but there is an appearance of instability. It is because the outline of the "housing" overhangs. It should be carried to the foundation without break in the continuity of the metal, as in Fig. 161.

130. When the support is divided up into several parts, modification of these principles becomes necessary, as the divisions require separate treatment. This question may be illustrated by

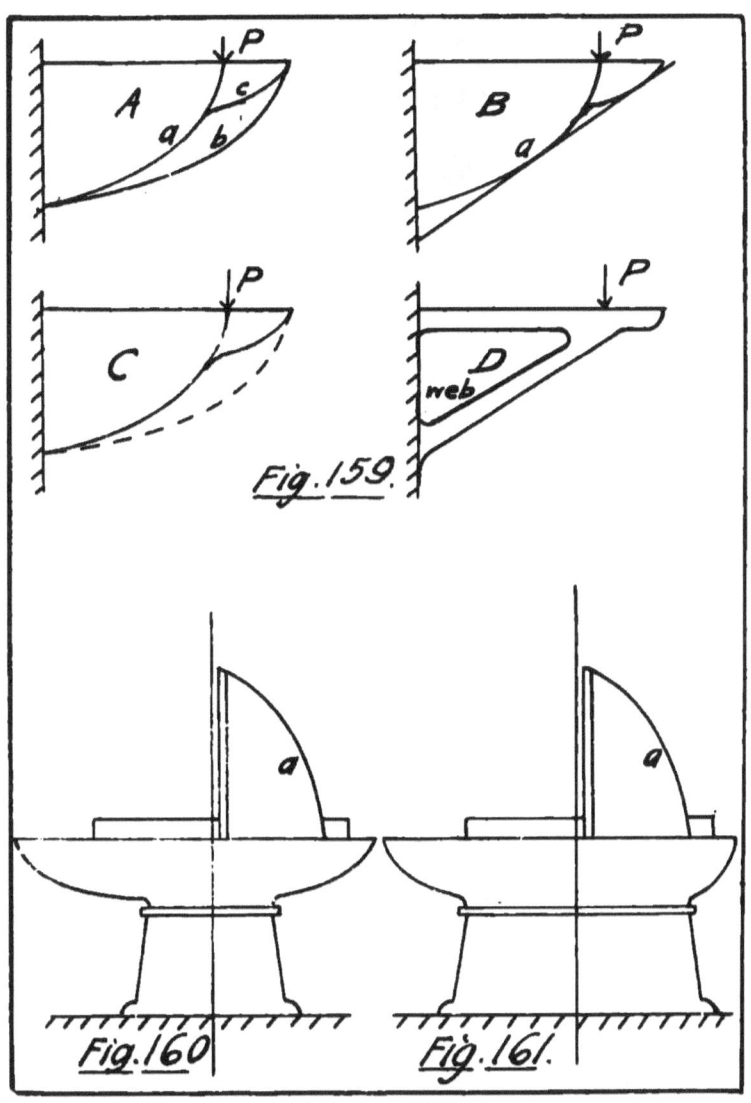

lathe supports. In Fig. 162 are shown three forms of support for a lathe, seen from the end. For stability the base needs to be broader than the bed. In A the width of base necessary is determined and the outlines are straight lines. The unnecessary material is cut away on the inside, leaving legs, which are compression members of correct form. The cross brace is left to check any tendency to buckle. For convenience to the workmen it is desirable to narrow this support somewhat without narrowing the base. The cross brace converts the single compressson member into two compression members. It is allowable to give these different angles with the vertical. This is done in B and the straight lines are blended into each other by a curve. C shows a common incorrect form of lathe support, the compression members from the cross brace downward being curved. There is no reason for this curved form. It is less capable of bearing its compressive load than if it were straight, and is no more stable than the form b; the width of base being the same.

Consider the lathe supports from the front. Four forms are shown in Fig. 163. If there were any force tending to move the bed of the lathe endwise the forms B and C would be allowable. But there is no force of this kind, and the correct form is the one shown in D. Carrying the foot out as in A, B, and C, increases the distance between supports (the bed being a beam with end supports and the load between); this increases the deflection and the fibre stress due to the load. This increase in stress is probably not of any serious importance, but the principle should be regarded or the appearance of the machine will not be right. If the supports were joined by a cross member, as in Fig. 164, they would be virtually converted into a single support, and should then taper from all sides.

131. If a machine be supported on a single box pillar, change in the form of the foundation cannot induce stress in the machine frame tending to change its form. If, however, the machine is supported on four or more legs the foundation might sink away from one or more of them and leave a part unsupported. This might cause torsional or flexure stress in some part of the machine, which might change its form, and interfere with the accuracy of its action.

But if the *machine be supported on three points* this cannot occur, because, if the foundation should sink under any one of the supports, the support would follow and the machine would still rest on three points. When it is possible, therefore, a machine which cannot be carried on a single pillar should be supported on three points. Many machines are too large for three-point support, and the resource is to make the bed, or part supported, of box section and so rigid that even if some of the legs should be left without foundation, the part supported would still maintain its form. More supports are often used than are necessary. Thus, if a lathe has two pairs of legs like those shown in *B*, Fig. 162, and these are bolted firmly to the bed, there will be four points of support. But if, as suggested by Professor Sweet, one of these pairs be connected to the bed by a pin so that the support and the bed are free to move, relatively to each other, about the pin, as in Fig. 165, then this is equivalent to a single support, and the bed will have three points of support, and will maintain its form independently of any change in the foundation. This is of special importance when the machines are to be placed upon yielding floors.

132. Fig. 166 shows another case in which the number of supports may be reduced without sacrifice. In *A* three pairs of legs are used. There are therefore six points of support. In *B* two pairs of legs are used and one may be connected by a pin, and there will be but three points of support. The chance of the bed being strained from changing foundation, has been reduced from 6 in *A* to 0 in *B*. The total length of bed is 12 ft., and the unsupported length is 6 ft. in both cases.

133. Figs. 167 and 168 show correct methods of support for small lathes and planers, due to Professor Sweet. In Fig. 167 the lathe "head stock" has its outlines carried to the foundation by the box pillar; *a* represents a pair of legs connected to the bed by a pin connection, and instead of being placed at the end of the bed it is moved in somewhat, the end of the bed being carried down to the support by a parabolic outline. The unsupported length of bed is thereby decreased, the stress on the bed is less, and the bed will

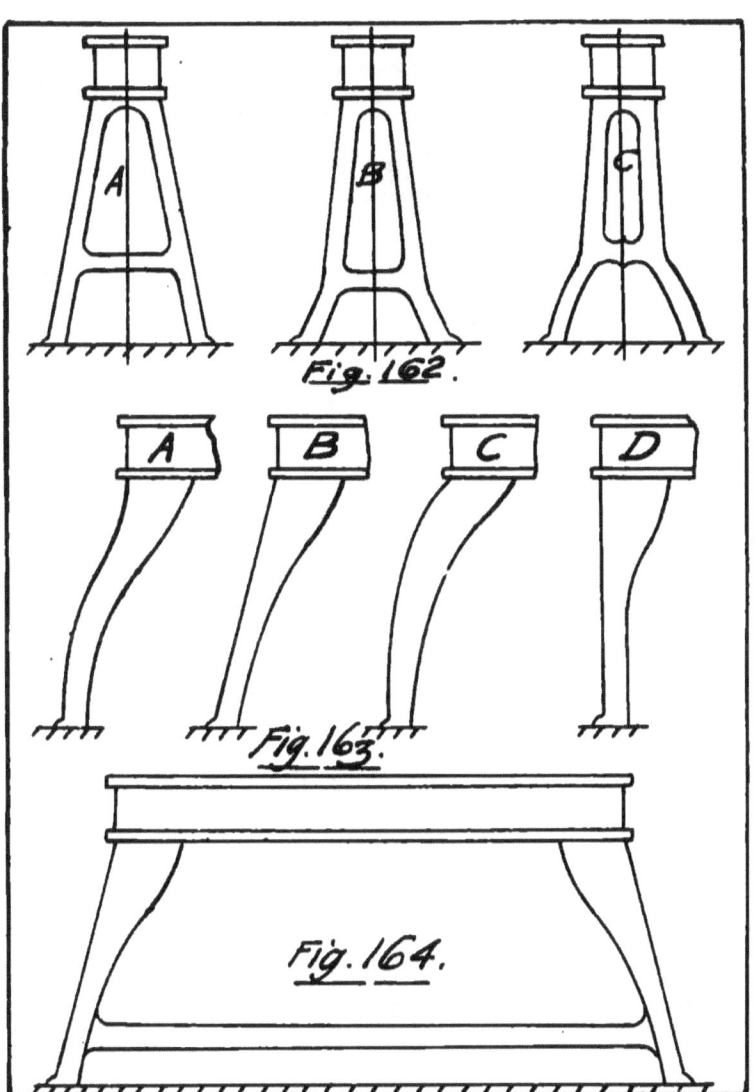

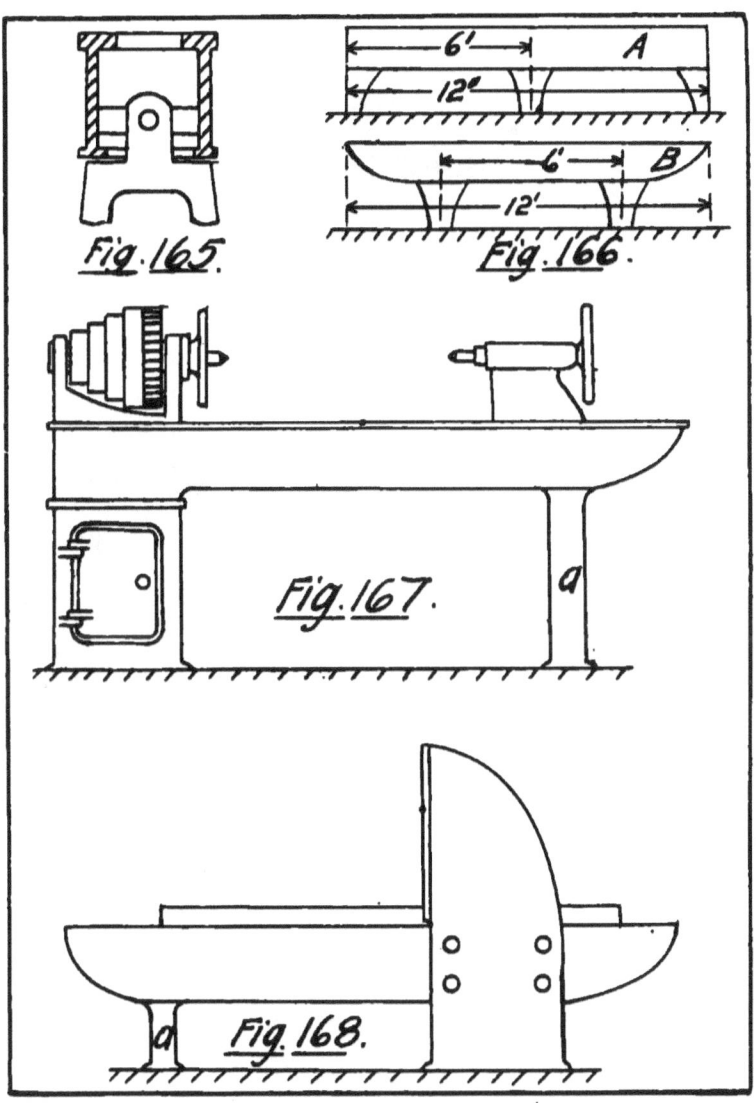

maintain its form regardless of any yielding of the floor or foundation. In Fig. 168 the housings, instead of resting on the bed as is usual in small planers, are carried to the foundation, forming two of the supports; the other is at a and has a pin connection with the bed, which being thus supported on three points cannot be twisted or flexed by a yielding foundation.

CHAPTER XVI.

MACHINE FRAMES.

134. Fig. 169 shows an **open side frame,** such as is used for punching and shearing machines. During the action of the punch or shear a force is applied to the frame tending to separate the jaws. This force may be represented in magnitude, direction, and line of action by P. It is required to find the resulting stresses in the three sections AB, CD, and EF. Consider AB. Let the portion above this section be taken as a free body. The force P, Fig. 170, and the opposing resistances to deformation of the material at the section AB, are in equilibrium. Let H be the projection of the gravity axis of the section AB, perpendicular to the paper. Two equal and opposite forces, P_1 and P_2, may be applied at H without disturbing the equilibrium. Let P_1 and P_2 be each equal to P, and let their line of action be parallel to that of P. The free body is now subjected to the action of an external couple, Pl, and an external force, P_1. The couple produces flexure about H, and the force P_1 produces tensile stress in the section AB. The flexure results in a tensile stress varying from a maximum value in the outer fibre at A to zero at H, and a compressive stress varying from a maximum in the outer fiber at B to zero at H. This may be shown graphically at JK. The ordinates of the line LM represent the varying stress due to flexure; while ordinates between LM and NO represent the uniform tensile stress. This latter diminishes the compressive stress at B, and increases the tensile stress at A. The tensile stress per square inch at A therefore equals $S + S_1$; where S equals the unit fibre stress due to flexure at A, and S_1 equals the unit tensile stress due to P_1. Now $S = \dfrac{Plc}{I}$, and $S_1 = \dfrac{P}{A}$; in which $c =$ the distance

MACHINE FRAMES. 151

from the gravity axis to the outer fibre $= AH$, and $I =$ the moment of inertia of the section about H, and $A =$ area of the cross-section AB.

Let it be required to design the frame of a machine to punch $\frac{3}{4}''$ holes in $\frac{1}{2}''$ steel plates, 18" from the edge. The surface resisting the shearing action of the punch $= \pi \times \frac{3}{4}'' \times \frac{1}{2}'' = 1.17$ sq. in. The ultimate shearing strength of the material is say 50000 pounds per square inch. The total force, P, which must be resisted by the punch frame $= 50000 \times 1.17 = 58500$ pounds.

135. The material and form for the frame must first be selected. The form is such that forged material is excluded, and difficulties of casting and high cost exclude steel casting. The material, therefore, must be cast iron. Often the same pattern is used both for the frame of a punch and shear. In the latter case when the shear blade begins and ends its cut the force is not applied in the middle plane of the frame, but considerably to one side, and a torsional stress results in the frame. Combined torsion and flexure are best resisted by members of box form. The frame will therefore be made of cast iron and of box section. The dimension AB may be assumed so that its proportion to the "reach" of the punch appears right; the width and thickness of the cross-section may also be assumed. From these data the maximum stress in the outer fibre may be determined. If this is a safe value for the material used the design will be right.

136. Let the assumed dimensions be as shown in Fig. 171. Then

$$A = b_1 d_1 - b_2 d_2 = 78 \text{ sq. in.}$$

$$I = \frac{b_1 d_1^3 - b_2 d_2^3}{12}$$

$= 8000$ bi-quadratic inches, nearly.

$c = d_1 \div 2 = 9''$; $l =$ the reach of the punch $+ c = 27''$; $P = 58500$ lbs., as determined above. Then

$$S_1 = \frac{P}{A} = \frac{58500}{78} = 770,$$

$$S = \frac{Plc}{I} = \frac{58500 \times 27 \times 9}{3000} = 4860.$$

$S_1 + S = 5630 =$ maximum stress in the section.

The average strength of cast iron such as is used for machinery castings, is about 20000 lbs. per square inch. The factor of safety in the case assumed equals $20000 \div 5630 = 3.5$. This is too small. There are two reasons why a large factor of safety should be used in this design: I. When the punch goes through the plate the yielding is sudden and a severe stress results. This stress has to be sustained by the frame, which for other reasons is made of unresilient material. II. Since the frame is of cast iron, there will necessarily be shrinkage stresses which the frame must sustain in addition to the stress due to external forces. These shrinkage stresses cannot be calculated and therefore can only be provided against by a large factor of safety.

Cast iron is strong to resist compression and weak to resist tension, and the maximum fibre stress is tension on the inner side. The metal can therefore be more satisfactorily distributed than in the assumed section, by being thickened where it sustains tension, as at a, Fig. 172. If, however, there is a very thick body of metal at a, sponginess and excessive shrinkage would result. The form B would be better, the metal being arranged for proper cooling and for the resisting of flexure stress.

137. Dimensions may be assigned to a section like B and the cross-section may be checked for strength as before. See Fig. 173. GG, a line through the centre of gravity of the section, is found to be at a distance of 7" from the tension side. The required values are as follows: $c = 7"$; $l =$ reach of punch $+ c = 18 + 7 = 25"$; $A = 156.5$ sq. in.; $I = 5000$ bi-quadratic inches, nearly; $P = 58500$ lbs.

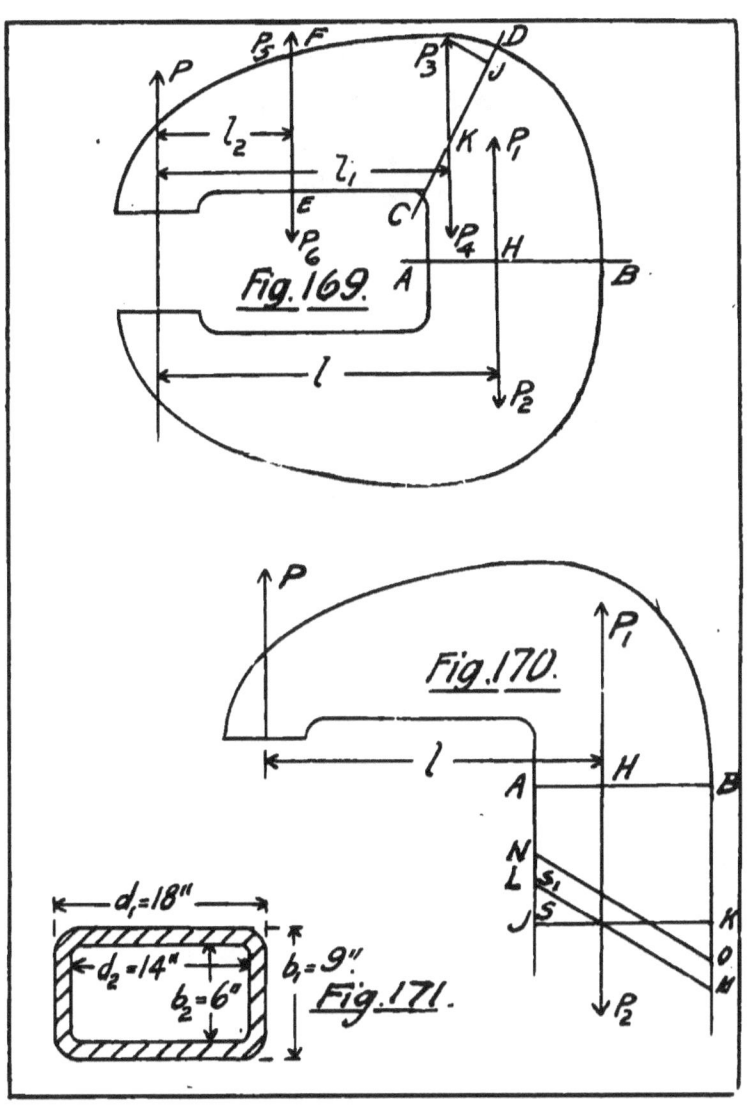

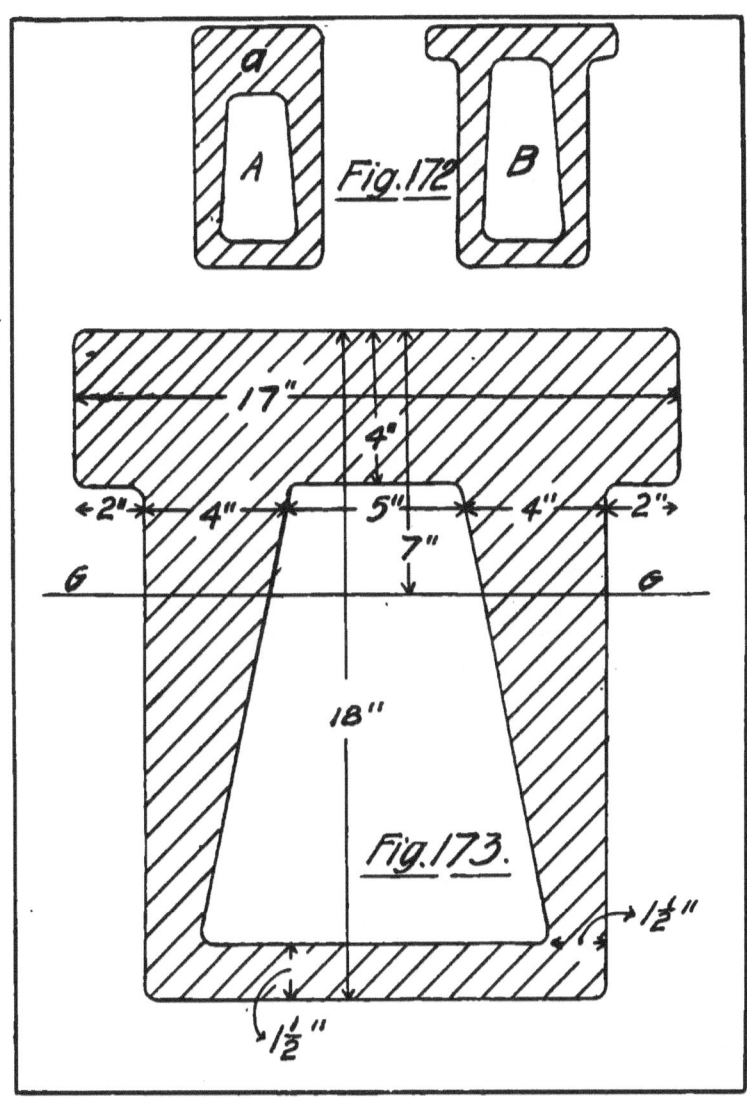

MACHINE FRAMES.

Then $\quad S_1 = \dfrac{P}{A} = \dfrac{58500}{156\cdot 5} = 374$ lbs.

$$S = \dfrac{Plc}{I} = \dfrac{58500 \times 25 \times 7}{5000} = 2047 \text{ lbs.}$$

$S_1 + S = 2421$ lbs. = maximum fibre stress in the section. The factor of safety $= 20000 \div 2421 = 8\cdot 25$. This section, therefore, fulfills the requirement for strength, and the material is well arranged for cooling with little shrinkage and without spongy spots. The gravity axis may be located, and the value of I determined by graphic methods. See Hoskins' "Graphic Statics."

138. Let the section CD, Fig. 169, be considered. Fig. 174 shows the part at the left of CD free. K is the projection of the gravity axis of the section. As before, put in two opposite forces P_3 and P_4, equal to each other and to P, and having their common line of action parallel to that of P, at a distance l_1 from it. P and P_4 now form a couple, whose moment $= Pl_1$, tending to produce flexure about K. P_3 must be resolved into two components, one $P_3 J$, at right angles to the section considered, tending to produce tensile stress; and the other JK, parallel to the section, tending to produce shearing stress. The greatest unit tensile stress in this section will equal the sum of that due to flexure and that due to tension =

$$S + S_1 = \dfrac{Plc}{I} + \dfrac{P_3 J}{A}.$$

The greatest unit shear $= \quad S_s = \dfrac{JK}{A}.$

139. In the section FE, Fig. 169, which is parallel to the line of action of P, equal and opposite forces, each $= P$, may be introduced, as P_5 and P_6. P and P_6 will then form a couple with an arm l_2, and P_5 will be wholly applied to produce shearing stress. The maximum unit tensile stress in this section will be that due to flexure, $S = Pl_2 c \div I$, and the maximum unit shear will be $S_s = P \div A$. Any section may be thus checked.

140. The dimensions of several sections being found, the outline curve bounding them should be drawn carefully, to give good appearance. The necessary modifications of the frame to provide for support, and for the constrainment of the actuating mechanism, may be worked out as in Fig. 175. A is the pinion on the pulley shaft from which the power is received; B is the gear on the main shaft; C, D, and G are parts of the frame added to supply bearings for the shafts; E furnishes the guiding surfaces for the punch "slide." The method of supporting the frame is shown, the support being cut under at F for convenience to the workman. The parts C, D, E, and G can only be located after the mechanism train has been designed.

141. **Slotting Machine Frame.**—See Fig. 176. It is specified that the slotter shall cut at a certain distance from the edge of any piece, and the dimension AH is thus determined. The table G must be held at a convenient height above the floor, and RK must provide for the required range of "feed." K is cut under for convenience to the workman, and carried to the floor line as shown. It is required to "slot" a piece of given vertical dimension, and the distance from the surface of the table to E is thus determined. Let the dimension LM be assumed so that it shall be in proper proportion to the necessary length and height of the machine. The curves LS and MT may be drawn for bounding lines of a box frame to support the mechanism. M should be carried to the floor line as shown, and *not cut under*. None of the part DNE, nor that which serves to support the cone and gears on the other side of the frame, should be made flush with the surface $LSTM$, because nothing should interfere with the continuity of the curves LS and TM. *The supporting frame of a machine should be clearly outlined, and other parts should appear as attachments.* The member VW should be designed so that its inner outline is nearly parallel to the outline of the cone pulley, and should be joined to the main frame by a curve. The outer outline should be such that the width of the member increases slightly from W to V, and should also be joined to the main frame by a curved outline. In any cross-section of the frame,

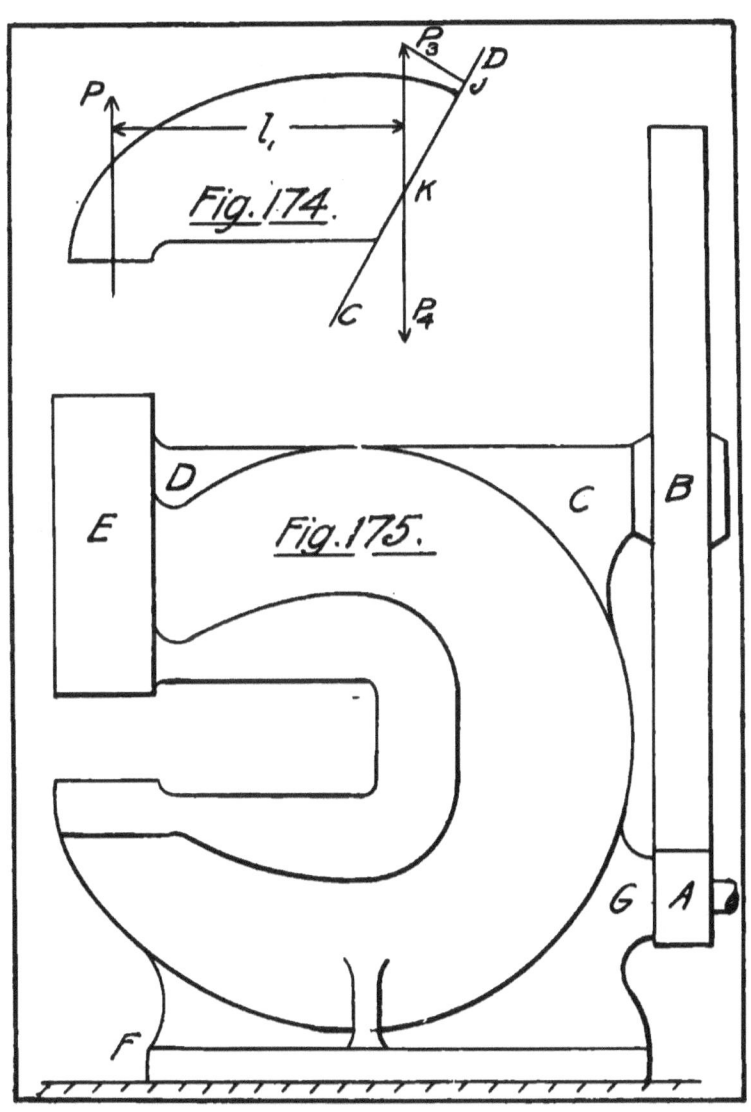

MACHINE FRAMES. 155

as XX, the amount of metal and its arrangement may be controlled by the core. It is dictated by the maximum force, P, which the tool can be required to sustain. The tool is carried by the slider of a slider crank chain. Its velocity varies, therefore, from a maximum near mid-stroke, to zero at the upper and lower ends of its stroke. The belt which actuates the mechanism runs on one of the steps of the cone pulley, at a constant velocity. Suppose that the tool is set (accidentally) so that it strikes the table just before the slider has reached the lower end of its stroke. The resistance, R, offered by the tool to being stopped, multiplied by its (very small) velocity, equals the difference of belt tension multiplied by the belt velocity (friction and inertia neglected). R, therefore, would vary inversely as the slider velocity, and hence may be very great. Its maximum value is indeterminate. A "breaking piece" may be put in between the tool and the crank. Then when R reaches a certain value, the breaking piece fails. The stress in the stress-members of the machine is thereby limited to a certain definite value. From this value the frame may be designed. Let $P =$ upward force against the tool when the breaking piece fails. Let $l =$ the horizontal distance from the line of action P to the gravity axis of the section XX. Then the section XX sustains flexure stress caused by the moment Pl, and tensile stress equal to P. The maximum unit stress in the section $=$

$$S + S_1 = \frac{Plc}{I} + \frac{P}{A}.$$

A section may be assumed and checked for safety, as for the punch.

142. Stresses in the Frame of a Side-Crank Steam Engine. — Fig. 177 is a sketch in plan of a side-crank engine of the "girder bed" type. The supports are under the cylinder C, the main bearing E, and the out-board bearing D. A force P is applied in the centre line of the cylinder, and acts alternately toward the right and toward the left. In the first case it tends to separate the cylinder and main shaft; and in the second case it tends to bring them nearer together. The frame resists these tendencies with resulting internal stresses.

Let the stresses in the section AB be considered. The end of the frame is shown enlarged in Fig. 178. If the pressure from the piston is toward the right, the stresses in AB will be: I. Flexure due to the moment Pl, resulting in tensile stress below the gravity axis N, with a maximum value at b, and a compressive stress above N with a maximum value at a. II. A direct tensile stress, $= P$, distributed over the entire section, resulting in a unit stress $= P \div A = S$ lbs. per sq. in. This is shown graphically at n, Fig. 178. $a_1 b_1$ is a datum line whose length equals AB. Tensions are laid off toward the right and compressions toward the left. The stress due to flexure varies directly as the distance from the neutral axis N_1, being zero at N_1. If, therefore, $b_1 c_1$ represents the tensile stress in the outer fibre, then $c_1 k_1$ drawn through N will be the locus of the ends of horizontal lines, drawn through all points of $a_1 b_1$, representing the intensity of stress, in all parts of the section, due to flexure. If $c_1 d_1$ represent the unit stress due to direct tension, then, since this is the same in all parts of the section, it will be represented by the horizontal distance between the parallel lines $c_1 k_1$ and $d_1 e_1$. This uniform tension increases the tension $b_1 c_1$ due to flexure, causing it to become $b_1 d_1$; and reduces the compression $k_1 a_1$, causing it to become $e_1 a_1$. The maximum stress in the section is therefore tensile stress in the lower outer fibre, and is equal to $b_1 d_1$.

When the force P is reversed, acting toward the left, the stresses in the section are as shown at m: compression due to flexure in the lower outer fibre equal to $c_2 b_2$; tension due to flexure in the upper outer fibre equal to $a_2 k_2$; and uniform compression over the entire surface equal to $d_2 c_2$. This latter increases the compression in the lower outer fibre from $b_2 c_2$ to $b_2 d_2$, and decreases the tension in the upper outer fibre from $a_2 k_2$ to $a_2 e_2$. The maximum stress in the section is therefore compression in the lower outer fibre equal to $b_2 d_2$. The maximum stress, therefore, is always in the side of the frame next to the connecting-rod.

If the gravity axis of the cross-section be moved toward the

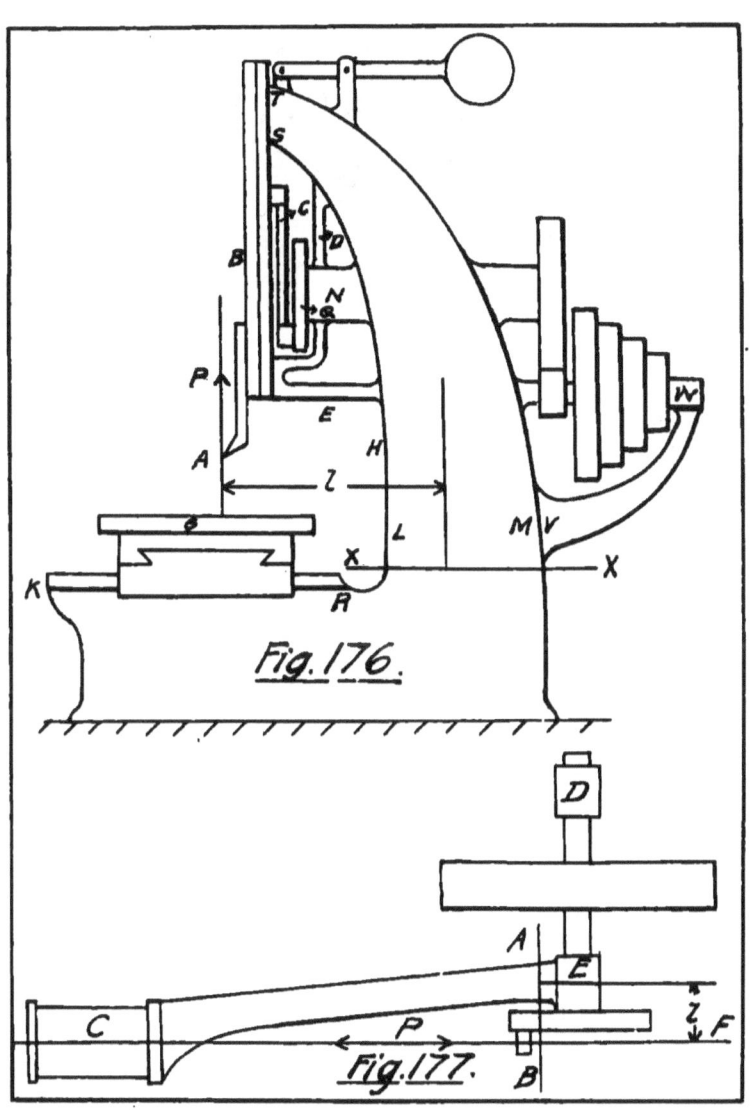

Fig. 176.

Fig. 177.

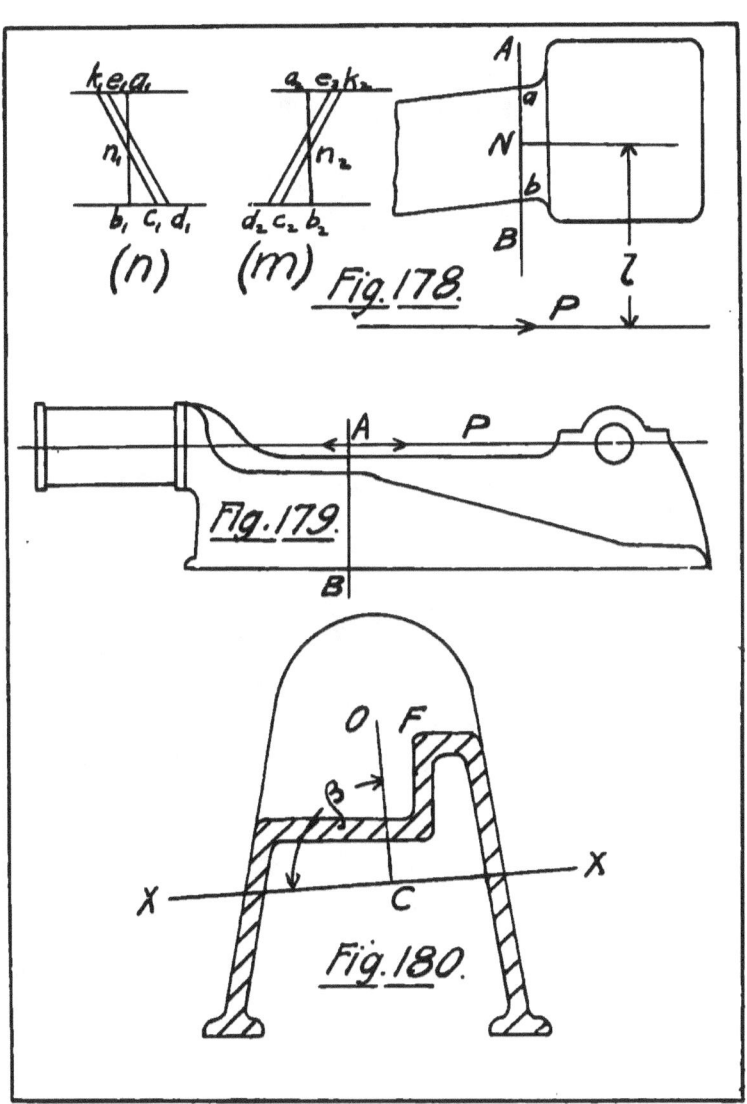

connecting-rod, the stress in the upper outer fibre will be increased, and that in the lower outer fibre will be proportionately decreased. The gravity axis may be moved toward the connecting-rod by increasing the amount of material in the lower part of the cross-section and decreasing it in the upper part.

The stresses in any other section nearer the cylinder will be due to the same force, P, as before; but the moment tending to produce flexure will be less, because the lever arm of the moment is less and the force constant.

143. Suppose the engine frame to be of the type which is continuous with the supporting part as shown in Fig. 179. Let Fig. 180 be a cross-section, say at AB. O is the centre of the cylinder. The force P is applied at this point perpendicular to the paper. C is the centre of gravity of the section (the intersection of two gravity axes perpendicular to each other, found graphically). Join C and O, and through C draw XX perpendicular to CO. Then XX is the gravity axis about which flexure will occur.* The dangerous stress will be at F, and the value of c will be the perpendicular distance from F to XX. The moment of inertia of the cross-section about XX may be found, $= I$; l, the lever arm of P, $= OC$. The stress at F, $S + S_1$, must be safe value.

$$S = \frac{Plc}{I}, \qquad \text{in known terms.}$$

$$S_1 = \frac{P}{\text{area of sec'n}}, \qquad \text{in known terms.}$$

* This is not strictly true. If OC is a diameter of the "ellipse of inertia," flexure will occur about its conjugate diameter. If the section of the engine frame is symmetrical with respect to a vertical axis, OC is vertical, and its conjugate diameter XX is horizontal. Flexure would occur about XX, and the angle between OC and XX would equal 90°. As the section departs from symmetry about a vertical, XX, at right angles to OC, departs from OC's conjugate, and hence does not represent the axis about which flexure occurs. In sections like Fig. 178, the error from making $\beta = 90°$ is unimportant. When the departure from symmetry is very great, however, OC's conjugate should be located and used as the axis about which flexure occurs. For method of drawing "ellipse of inertia" see Hoskins' "Graphic Statics."

144. Closed Frames.— Fig. 181 shows a closed frame. The members G and H are bolted rigidly to a cylinder C at the top, and to a bed plate, DD, at the bottom. A force P may act in the centre line, either to separate D and C, or to bring them nearer together. The problem is to design G, H, and D for strength. If the three members were "pin connected," see Fig. 182, the reactions of C upon A and B at the pins would act in the lines EF and GH. Then if P acts to bring D and C nearer together, compression results in A, the line of action being EF; compression results in B, the line of action being GH. These compressions being in equilibrium with the force P, their magnitude may be found by the triangle of forces. From these values A and B may be designed. C is equivalent to a beam whose length is l, supported at both ends, sustaining a transverse load P, and tension equal to the horizontal component of the compression in A or B. The data for its design would therefore be available. Reversing the direction of P reverses the stresses; the compression in A and B becomes tension; the flexure moment tends to bend C convex downward instead of upward, and the tension in C becomes compression.

145. But when the members are bolted rigidly together, as in Fig. 181, the lines of the reactions are indeterminate. Assumptions must therefore be made. Suppose that G is attached to D by bolts at E and A. Suppose the bolts to have worked slightly loose, and that P tends to bring C and D nearer together. There would be a tendency, if the frame yields at all, to relieve pressure at E and to concentrate it at A. The line of the reaction would pass through A and might be assumed to be perpendicular to the surface AE. Suppose that P is reversed and that the bolts at A are loosened, while those at E are tight. The line of the reaction would pass through E, and might be assumed to be perpendicular to EA. MN is therefore the assumed line of the reaction, and the intensity $= P \div 2$. In any section of G, as XX, let K be the projection of the gravity axis. Introduce at K, two equal and opposite forces, equal to R and with their lines of action parallel to that of R. Then in the section there is flexure stress due to the flexure

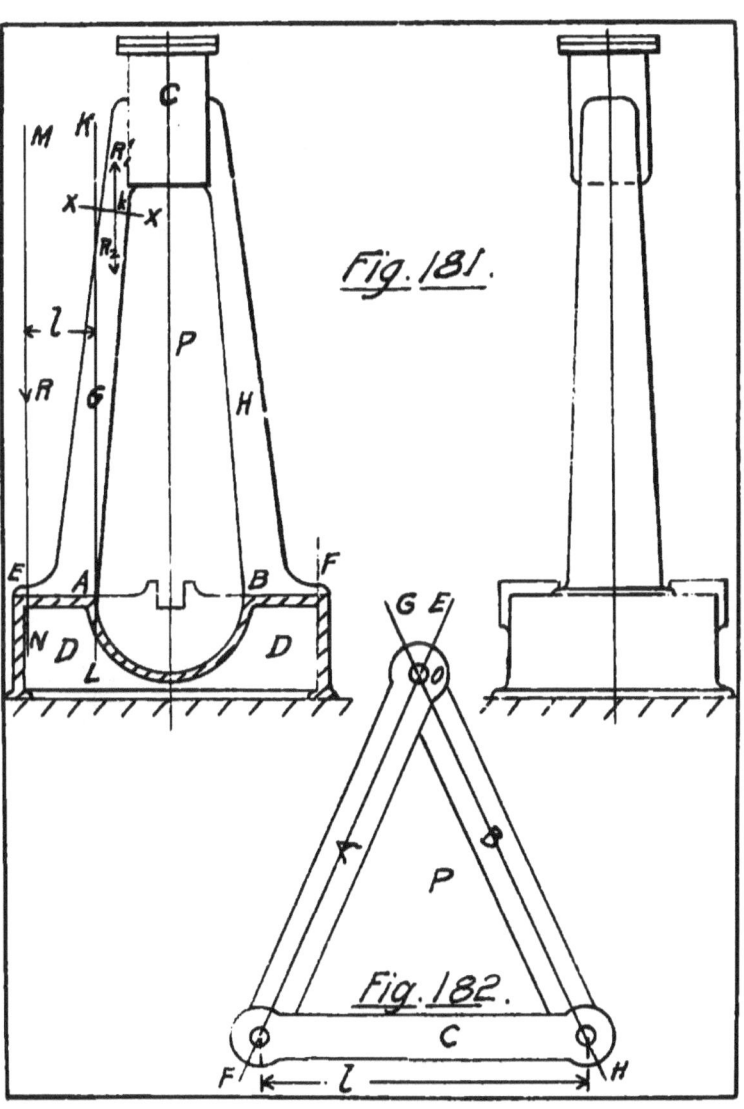

Fig. 181.

Fig. 182.

MACHINE FRAMES. 159

moment Rl, and tensile stress due to the component of R_2 perpendicular to the section, $= R_3$. Then the maximum stress in the section $= S + S_1$.

$$S = \frac{R_3}{A}; \quad S_1 = \frac{Rlc}{I}.$$

A section may be assumed, and A, I, and c become known; the maximum stress also becomes known, and may be compared with the ultimate strength of the material used.

Obviously this resulting maximum stress is greater when the line of the reaction is MN, than when it is KL. Also it is greater when MN is perpendicular to EA, than if it were inclined more toward the centre line of the frame. The assumptions therefore give safety. If the force P could only act downward, as in a steam hammer, KL would be used as the line of the reaction.

146. The part D in the bolted frame, is not equivalent to a beam with end supports and a central load like C, Fig. 182, but more nearly a beam built in at the ends with central load; and it may be so considered, letting the length of the beam equal the horizontal distance from E to F, $= l$. Then the stress in the mid-section will be due to the flexure moment $\frac{Pl}{8}$, and the maximum stress $= S = \frac{Plc}{8I}$. The values c and I may be found for an assumed section, and S becomes known.

INDEX.

Addendum, 148.
Angularity of connecting-rod, 20.
Annular gears, 50.

Belts, 75.
   centrifugal force of, 89.
   design of, 81.
   transmission by, 75.
Bevel gears, 59.
Bolts and screws, 130.
   and screws, design of, 131.
   cross-section of, to resist tension, 135.
   elongation of, 134.
   fastenings to hold steam chest cover, 132.
   tendency to loosen nuts, 136.

Cams, 72.
Cantilever in machines, 145.
Centro of two gears, 40.
Centros of relative motion, 13.
   in compound mechanism, 14.
Complete constrainment of motion, 5.
Cone pulleys, 78.
Conservation of energy, 1.
Constrained motion, 3.
Cotter, 138.
Cycloidal curves, 42.
   teeth, 60.

Energy in machines, 28.
   is transferred in time, 29.

Feathers, or splines, 137.
   rules for, 137.
Fly-wheel, pump, 97.
   steam engine, 99.
   design of, 93.
Force problems, 29.
   in the steam engine, 35
Form of machine parts, 140.
   in stress, tension, or compression, 140.
   in transverse stress, 141.
   in torsional stress, 142.
Frame of machine, 143.
   slotting machine, 154.
   closed, 158.
Free motion, 3.
Function of machines, 2.

Gears: angular velocity ratio, 70.
   annular, 50.
   backlash, 48.
   bevel, 59.
   bevel, design of, 63.
   clearance, 48.
   compound spur gear chains, 69.
   definitions, 45, 47.
   design of worm, 67.

Gears, diametral pitch, 48.
 face, 48.
 formulas, 55.
 interchangeable, 51.
 interchangeable involute, 53.
 laying out, 54.
 skew bevel, 64.
 solution from other data, 68.
 strength of teeth, 56.
 total depth, 48.
 tooth, design of, 57.
 spiral, 64.
 working depth, 48.
 worm, 65.

Generating circle, 43.
Guides, 128.

Higher pairs, 39.

Independent constrainment of motion, 13.
Instantaneous motion and instantaneous centres or centros, 7.
Interchangeable gears, 43.
Involute teeth, 61.
 tooth outlines, 46.

Journals, design of, 111.
 allowable pressure, 112.
 bearings and boxes, 120.
 direction of motion, 112.
 frictional resistance of, 113.
 lubrication of, 123.
 oiling, 124.
 pressure in thrust, 118.
 radiation of, 114.
 stationary, 124.
 thrust, 117.

Keys, as a means for preventing relative rotation, 137.

Keys, rules for, 138.
Kinds of motion in machines, 6.

Lever crank chain, 13.
Linkages or motion chains; mechanisms, 11.
Location of centros, 12.
Loci of centros or centroids, 9.

Machine frames, 150.
Machine supports, 147.
Machinery of application, 2.
Machinery of transmission, 2.
Means for preventing relative rotation, 137.
Motion independent of force, 6.

Non-circular wheels, 59.

Open frame design, 152.
 side frames, 150.
Outline of machine frame, 154.

Pairs of motion elements, 10.
Parallel or straight line motions, 37.
Passive resistance, 3.
Pitch point, 41.
Prime mover, 2.
Pulleys, cone, graphical method, 79.

Racks, 48.
Rate of doing work, 28.
Ratio of a quick return, 21.
Reduction of the number of supports, 148.
Relative linear velocity in same link, 17.
 linear velocity not in same link, 18.
 motion, 6.
Rigid body, 7.
Riveted joints, 100.

Riveted joints, butt, 101.
    lap, 101.
    margin of rivets, 106.
    pitch of rivets, 105.
    table for, 103.
    table for rivet diameters, 104.
    table of efficiency, 107.

Shrink and force fits, 139.
Slider crank chain, 12.
    mechanism, 12.
    and guide of unequal length, 127.
Sliding surfaces, 126.
Slotted cross-head mechanism, 14.
Slotting machine frame, 154.
Solution of a quick return, 23.
Steam engine, stresses in, 155.

Stresses in the frame of a steam engine, 155.
Support divided into several parts, 146.
    for lathes, 148.
    single box pillar, 147.

Table for use in designing belts, 85.
Tooth outlines, 41
Toothed wheels, or gears, 39.

Vector, 16.
Velocity, 15.
    of cutting tools, 21.

Whitworth quick return mechanism, 25.

www.ingramcontent.com/pod-product-compliance
Lightning Source LLC
Chambersburg PA
CBHW030820230426
43667CB00008B/1298